THE
JOHN WILLIAMS
STORY

Phœnix Iron Works
RHUDDLAN, NEAR RHYL
NORTH WALES

Gwynfor Williams

The Story of a Welsh Presbyterian Lay Preacher, Agricultural Machinery and Implement Maker.

First published 2016

Copyright © Gwynfor Williams 2016

ISBN: 978-0-9954689-0-0

A catalogue record of this book is available from the British Library.

Published by Gwynfor Williams, 32, Dorchester Road, Garstang, Preston. PR3 1HH

Printed in Wales by WPG (Welshpool Printing Group) Powys.

MR JOHN WILLIAMS

Contents

Acknowledgements

Probably the most comprehensive history of John Williams and Son Phoenix Iron Works Rhuddlan was written by the late Dr Elfyn Scourfield M.A. At this time Dr Scourfield was a Curator at The Welsh Folk Museum St Ffagans.

The fruits of his research are documented in the Flintshire Historical Society Journal Volume 28 1977–1978 and is available (Welsh Journals Online) National Library of Wales Aberystwyth. On his gleaning trips to North Wales Dr Scourfield visited our farm at Pwllglas and ignited my enthusiasm for Welsh Agricultural Machinery.

Special thanks must also go to Mr and Mrs John Dod of St Asaph and to their relative (the late) Frank Jones of Southport who loaned me several photographs. Frank also enabled me to piece together the career of his father Iorwerth who was joint works foreman at Rhuddlan. To Ted Reece of Bodelwyddan for information and photographs with reference to Part 3 The Ruddlan Foundry 1923-2000.

My thanks also to: The late John Bumby Penygroes, Gareth Beech Welsh Folk Museum St Ffagans, Cefnmeiriadog Historical Society, Reg Davies Rhuddlan, Denbighshire Record Office, Twm Elias (Fferm a Thyddyn) Emrys Evans Trawsfynydd, Flintshire Record Office, Stuart Gibbard Moulton Chapel Spalding, Greenfield Valley Trust Holywell, Lorna Jenner Alyn Books, Hans Jensen Benala Victoria Australia, Eifion Ll. Jones Clawddnewydd, Gareth Jones Clawddnewydd, Hefin L Jones St Asaph, Merfyn Jones (Lelo Metals) Paul Jones Betws yn Rhos, Myerscough College Library (to John Humphrey and staff), John Paterson Dyserth, Rhuddlan Local History Society, (the late Brian G. Roberts of Lancaster (Llanuwchllyn), Jo Roberts Llanrwst, The late Dr Elfyn Scourfield of Welsh Folk Museum St Ffagans, Karl Schroder Eglwys Cross, Whichurch, Ruth Smith Rhuddlan, Stewart Thomas Cresselly Pembs, Geraint O Williams Llanrhaeadr Denbigh, Brenda Wyatt Kinmel Bay.

Thanks to the Museum of English Rural Life, University of Reading. To Jennifer Glanville Assistant Curator and Caroline Benson Photographic Assistant. To John Hasings Garstang for cover designs and digital enhancement of photographs.

Last but not least
To Aeron Jones (compositor) Llanuwchllyn for his invaluable support and guidance.
Finally to the management and staff of WPG (Welshpool Printing Group) Powys.

Author

Gwynfor Williams was born on the farm of Llannerchgron Ucha Pwllglas Ruthin and is the author of the highly acclaimed book "The Jones Baler Story".

He is highly regarded for his knowledge of Welsh made agricultural machinery.

Educated at Llanelidan Primary and is a former pupil of Brynhyfryd Grammar School Ruthin.

On leaving school he attended Bangor Normal College and took up a teaching post at Mold. However, machinery was in the blood and after gaining engineering qualifications and experience at the Welsh College of Horticulture Northop moved to Garstang, Lancashire.

He is an honorary member of the Association of Lecturers in Agricultural Machinery and sings bass with Elswick Singers.

A retired Lecturer in Agricultural Machinery, he now turns the clock back to tell the "Story of John/Corbett-Williams and Son Phoenix Iron Works Rhuddlan Nr Rhyl".

Preface

As I write we are about to remember those who gave their lives in the First Word War 1914-18. It was at this time a hundred years ago that the premises of Corbett-Williams & Son Rhuddlan came under government control. These old premises formerly John Williams & Son on the banks of the river Clwyd had, for four decades, made some of Britain's finest agricultural machinery and exported all over the globe.

As more and more of the old foundry complex is re-developed any evidence as to its past will soon be lost.

The inspiration for this book has been my childhood experiences with the John Williams & Son chaff cutter (Chapter 3).

Foreword

This book is the culmination of a shared interest and enthusiasm for the work of John Williams and son, Phoenix Ironworks, Rhuddlan, and an awareness of the company's significance in Welsh agricultural machinery manufacture. Stimulated by collections and research at the then Welsh Folk Museum, St Fagans, it will substantially enhance the knowledge and interpretation of the agricultural machines and implements by Williams held in the collections of what is now St Fagans National History Museum.

Gwynfor Williams was inspired by an article titled *John Williams and son, Phoenix iron works, Rhuddlan,* by Dr Elfyn Scourfield, Keeper of the Department of Farming and Crafts, Welsh Folk Museum, published in 1977. The article had emanated from the acquisition of items made by Williams during the collecting of agricultural implements, machinery and tools from the 1960s onwards for the National Museum of Wales, and in particular for the new Agricultural Gallery, opened in 1974. The approach in museums during that period was to focus principally on agricultural technology and processes. As the open-air museum expanded, the re-erection of historical agricultural buildings and craft workshops, and organising exhibitions, were to take precedence over further research work.

Already interested in historical agricultural machinery of Welsh manufacture, Gwynfor had also recognised that there was much more information about Williams to be gathered, and machines studied and collected, to establish and comprehend the full picture. His extensive and very thorough research comprised primary research, as far afield as Australia, and collating documents, photos and illustrations from museums and libraries in Wales and beyond, as well as from private sources, into the most substantial single archive.

This has now been used to write a comprehensive history of John Williams, Phoenix Iron works, how it developed into Corbett-Williams, and the company's demise. It also charts how a successor company continued foundry work in Rhuddlan throughout the twentieth century. In addition to giving fulsome attention to the machines and their technological features, it also includes the lives of key personalities, the nature of Welsh agriculture in the nineteenth century, the effects of a changing labour market, and of emigration and new markets across the globe. The book includes a rich variety of photographs and illustrations, many from original company catalogues.

Integral to the story are several individuals, their characters, influences, and experiences: John Williams, the firm's founder and proprietor, also a Calvanistic Methodist minister for over sixty years, prominent in public life as one of the 'Fathers of Rhyl'; the highly-talented engineer John Whitaker, whose experience and expertise radically changed the company's output from casting railings and troughs to manufacturing a range of agricultural machines; and Francis Corbett, who re-invigorated the flagging company and exported new designs worldwide.

Before his passing, Elfyn Scourfield had expressed his delight at seeing the publication of *The Jones Balers Story,* Gwynfor's seminal book of the history of the renowned Flintshire machinery company, having built up a collection of their balers and other machines at St Fagans. John Williams's company was another significant Welsh manufacturer that had long needed to be fully chronicled, and its achievements understood, which have now been realised by this book.

Gareth Beech
Senior Curator: Rural Economy
St Fagans National History Museum
Amgueddfa Cymru – National Museum Wales

John Williams/Corbett -Williams & Son Timeline

1812 John Whitaker born at Little Mitton near Clitheroe, Lancashire.

1815 John Williams born at Dolwen near Abergele Denbighshire.

1836 John Williams establishes a Drapery business in Rhyl.

1847 Rhyl Railway Station opens.

1852 Elected Member of the First Board of Commissioners for Rhyl.

1858 Williams takes the lease of the Phoenix Ironworks at Rhuddlan.

1861 John Williams has an ironmongery business in Rhyl.

1869 John Whitaker becomes Head Design Engineer at Rhuddlan.

1871 Peter Williams (son) runs ironmongery business at 56 High Street, Rhyl.

1872 Cardiff Royal Show. Stand No 61 49 items exhibited including 14 chaff-cutters.

1873 Hull Royal Show

1874 Bedford Royal Show

1875 Taunton Royal Show "Princess" two horse mower trialled.

1876 Birmingham Royal Show.

1877 Liverpool Royal Show

1877 John Whitaker moves to form Powell Brothers and Whitaker Wrexham. John Williams joined by his son William becomes John Williams & Son.

1881 Death of Eunice Williams, wife of John Williams, aged 67 years.

1884 Shrewsbury Royal Show. Forty six John Williams machines exhibited.

1895 William Bridge Williams (son) takes over from his father.

1897 John Williams Testimonial Fund.

1898 Enforcement of Chaff-cutting Machines Accident Act.

1899 Death of John Williams aged 83 years.

1908 December. Corbett-Williams & Son Ltd.

1910 Major investment at Rhuddlan, exports worldwide.

1912 William Bridge Williams dies aged 62 years.

1914-18 War work with subsequent loss of exports.

1923 Corbett- Williams closes.

1923-2000 Rhuddlan Foundry.

2011 May demolished.

Background 1800-1900

The aim of this short chapter is to familiarize the reader with the world of John Williams born 1815 in rural North Wales.

The last decade of the eighteenth century saw the rise of Nonconformity and the rural Welsh population embraced Methodism. The ethos of Nonconformity was temperance and self-reliance. Sunday school would have enabled the young John Williams to discuss abstract ideas and become literate in Welsh. This was a time when chapels were being built at the rate of one a week.

At this time the population of Wales was less than one million but in two generations it would double.

Until 1830, agriculture remained the mainstay of the majority of the people of Wales, but farming was about to change from a hereditary custom to the practice of science.

Agricultural improvements received extensive coverage in the periodicals and a quarter of the space of the North Wales Gazette was devoted to them.

By 1770 farmers used lime on the land and rotated their crops with turnips and potatoes. The landowners fostered the establishment of county agricultural societies from which spawned county shows and agricultural colleges.

The development of the slate and coal industry allied with the coming of the railways saw thousands of farm labourers leaving the land.

By 1851 only 35% of the male population of Wales worked on the land. This figure by 1911 was to fall to 12%. An ever increasing population meant that agriculture had to feed more mouths with fewer workers. Change was inevitable and the country had to import food or raise home production. In Wales alone there would have been about forty thousand holdings employing two hundred thousand people.

As farmers turned to mechanization John Williams and Son would prosper.

John Williams & Son Stand at the Royal Show circa 1900.
Credit: Flintshire Record Office Reference PH/54/30.

Part 1
John Williams
1858-1908

Chapter 1

PERSONALITIES

The life of John Williams 1815-1899

Draper, Calvanistic Methodist Lay Preacher, Councillor, Businessman and Agricultural Engineer.

The Family

John Williams was born in 1815, in the hamlet of Dolwen Betws-yn-Rhos near Abergele, the eldest son of Peter R. Williams and Mary Bridge. In 1836, aged twenty-one, he established a drapery business at fifty-six High Street Rhyl living above the premises.

He married Eunice Roberts the daughter of the Reverend Peter Roberts the first minister of the Henry Rees Memorial Chapel Llansannan. The brother of Eunice was the bard Iorwerth Glan Aled (Edward Roberts 1819-1867).

In the year 1840 he commenced to preach and delivered his first sermon in Rhyl. A staunch Presbyterian, Williams would have travelled on horseback visiting chapels in an area consisting of most of the county of Flintshire, the Vale of Clwyd and the upland of Hiraethog between the Conway and Clwyd. After the evening service he would stay overnight (llety pregethwr) returning to Rhyl the following morning. It was during these preaching engagements that Williams observed the drudgery of farm work and the need for better implements to improve matters.

Clwyd Street Chapel in Rhyl was opened in 1855 but had a large debt. Williams was preaching at Clwyd Street and announced that if the collection came to £10 he would double it. He said "if it is one halfpenny less you will not have the £10". The collection came to £36 3s 4d and three farthings so he went to his pocket. The following year he set another challenge "if the collection reaches £70 I will make it up to £100" and again it worked.

MR JOHN WILLIAMS

John Williams

High Street Rhyl circa 1880
John Williams wearing top hat

In 1848 with the opening of the Bangor to Chester railway line the population of Rhyl rose dramatically and so did the drapery trade at fifty-six High Street.

John Williams as a successful businessman was elected to serve on the first Rhyl Board of Commissioners on its formation in 1852 and continued to serve until 1894. The Board consisted of thirty "men of impeccable character" and oversaw such works as water sewage and the building of the Pier.

Williams firmly believed in the future of Rhyl as a resort and built several houses and shops on the corner of High Street and Wellington Road. According to Slater's directory in1858 John Williams was the proprietor of the Phoenix Ironworks at Rhuddlan but more about that later.

At the Williams household between 1939 and 1851 seven children: four girls and three boys were born.

The 1861 Census for Fifty-Six high Street Rhyl indicates that Williams had an interest in ironmongery and perhaps that half his premises was given over to this business. In the following census seventeen persons were recorded at the Williams household.

Head – John Williams aged 46 Draper Calvanistic Minister
Wife – Eunice Williams aged 46
Children
      Mary A. aged 21 Draper's Assistant
      Elizabeth aged 19 Drapers' Assistant
      Isabella aged 18 Draper's Assistant
      Peter R. aged 16 Ironmonger's Assistant
      John B. aged 13 Scholar
      William B. aged 10 Scholar
      Sarah Letticia aged 5 Scholar

Four Males – 3 Ironmonger's Assistants   1 Draper's Assistant
Four Females – 2 Draper's Assistants   2 Domestic Servants.

1871 Census 91 Wellington Road

It is interesting to note it appears that Peter Roberts Williams, the son of John and Eunice, is now an Ironmonger at this address. Peter died in 1874 a young man aged 29 leaving a widow Annie and three children, Isabella, John and Sarah.

In 1881 they reside at 17 Wellington Road Tudor Place.

According to the 1881 Census of Elwy Villa 21 Elwy Street Rhyl

Residing on the evening of the census (a Sunday) were:

Eunice Williams aged 67 (dies on December 29th)

William Bridge Williams son aged 30 Agricultural Engineer and Machinery Maker.
Sarah L. Williams.

John Williams was on a preaching engagement and is registered at the home of Mr Mrs Thomas Grocers, of Chester Street Flint.

The original house and business **56 High Street** is now occupied by John B. Williams Draper (son of John and Eunice) his wife Mary and three children John H., Henrietta L. and William R.

John Williams continued as Principal at the Phoenix Ironworks and kept up his social activities within the community. When the National Eisteddfod of Wales visited Rhyl in 1894 his name appeared on the list of guarantors of the General Committee.

John Williams Agricultural Engineer

I wonder if before buying the Phoenix Iron Works (circa 1858), John Williams had done his version of "market research". It would have been interesting to have seen his "business plan".

The Phoenix Iron Works probably owes its origins to the shipbuilding industry of Rhuddlan and the Foryd. The opening of the Vale of Clwyd Railway rang the death knell for Rhuddlan's maritime industry and its dependent iron works. Skilled shipwrights, pattern-makers, moulders and blacksmiths would be now be idle.

Far from seeing this as doom and gloom, John Williams saw this as an opportunity to transform the Phoenix Iron Works into a thriving maker of agricultural machinery.

The newly opened Rhuddlan railway station less than a quarter of a mile away would be ideal for the supply of raw materials and export of finished products.

Since the railway had come to Rhyl there had been an explosion of development with all kinds of cast iron goods such as grates, stoves, mangling and wringing machines and lawn mowers arriving at the railway station.

Nearly everyone kept a pig and a market for cast iron feeding troughs was identified.

As the industrial revolution swung into gear, men were leaving the land for better wages in the mines and factories.

The chilled cast iron plough share seed drill reaper and portable steam thresher were now being produced in numbers. Market towns on the fertile soils of mainly Eastern England and Scotland became centres for machinery manufacturing ingenious blacksmiths became Agricultural Engineers. Farmers and owners of the larger farms with well drained flat fields embraced mechanization.

The Royal Agricultural Society of England set up trials of machinery so that the farmer could see and evaluate the machines on offer. The results of these trials were published in the annual RASE Journal. All major Agricultural Show Societies gave medals and awards.

Wales being a predominantly a stock raising area with wet soils and steep land meant that mechanization was slow; the rural population was still high, with women and children expected to work on the farm especially at harvest time.

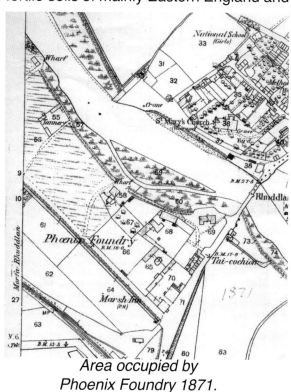

Area occupied by Phoenix Foundry 1871.

John Whitaker 1812–1896 Master Iron Founder and Design Engineer

The engineering genius John Whitaker combined with the business character of John Williams spells success. The foundry is transformed from a producer of gates and railings, pig troughs and ornamental goods to the manufacturer of the finest farm machinery.

Whitaker moves to Rhuddlan

1871 Census Town of Rhuddlan. No12 High Street:
Head John Whitaker aged 69 Engineer (John Williams and Son)
Wife Sarah Whitaker aged 62
George Whitaker aged 20 Bookkeeper (John Williams and Son)

It is not known when, how, or where John Williams and John Whitaker met. It is likely that Williams was actively seeking a man with patents granted in his name and would be making large cash offers. A man of the calibre of Whitaker with patents in the field of "barn machinery" could command good money, in the region of £400-£500 for his rights of manufacture.

Career

Circa 1850-1870 Picksley and Sims and Co Bedford Foundry Leigh Lancashire.
Circa 1870-1877 John Williams Phoenix
Iron Works Rhuddlan Near Rhyl North Wales.
1877 Powell Brothers and Whitaker Cambrian Iron Works Wrexham.

John Whitaker was born in the village of Little Mitton near Clitheroe in Lancashire.

According to the 1861 Census for the township of Leigh, he, his wife Sarah and six children lived at No 19 Bradshawgate in the township of Pennington Flash.

1861 Census Township of Pennington Flash. No19 Bradshawgate.

Head John Whitaker aged 49 years Master Iron Founder b Little Mitton.
Wife Sarah aged 52 b Dutton Lancashire.
Children
William aged 23 Mechanic (all branches)   b Mitton.
John aged 21 Iron Turner.   "
Ellen aged 19 Housemaid.   "
Richard aged 15 Mechanic (all branches)   "
Elizabeth aged 13
George aged 10 Scholar.

Picksley and Sims founded in 1845

In 1862 Picksley and Sims produced a 32 page catalogue of their Agricultural Implements confirming them as one of Britain's leading manufacturers.

As designer for the company John Whitaker is accredited with patents regarding chaff cutters and root pulpers.

Picksley and Sims also made several machines patented by other inventors an example is "Bamlett's Patent Reaper" judged the best reaper at the RASE Trial of Reaping Machines held at Garforth Leeds on August the 26th 1861.

John Whitaker's wife Sarah died aged 67 years on June 2nd 1876 and was buried at Rhuddlan church cemetery. Several factors may have influenced John Whitaker's decision to leave Rhuddlan for Wrexham.
John Williams had a son (William Bridge Williams) who was groomed to run the business.

The Powell Brothers Cambrian Iron Works Wrexham were young men with big ideas and offered John Whitaker the sum of £500 for patents and a partnership in their company; in 1877 Powell Brothers and Whitaker formed. Powell Brothers also offered employment to the Whitaker boys, William and George.

John Whitaker died aged 85 years on July13th 1896 and was buried with his wife Sarah at Rhuddlan. William Whitaker takes over the partnership at Powell.

Mower seat

POWELL BROTHERS
AND WHITAKER,
Cambrian Iron Works,
WREXHAM.

Powell

Part 1
John Williams
1858-1908

<u>Chapter 2</u>
IRONMONGERY AND IMPLEMENTS

No. 10.

CATALOGUE

OF AGRICULTURAL

IMPLEMENTS,

MACHINES,

ETC.,

MANUFACTURED BY

JOHN WILLIAMS,

ENGINEER

AGRICULTURAL MACHINIST, &c.,

PHŒNIX IRON WORKS,

RHUDDLAN, NEAR RHYL.

March, 1875.

Miscellaneous Products

Entrance Gates, Palisading, Tomb and Balcony Railing, Gas Lamp Standards

This was a time of great chapel building. John Williams with his "preaching connections" would have profited.

The gates and railings of Bathafarn Wesleyan Methodist Chapel on Market Street Ruthin bear testimony to the workmanship of John Williams. Built in 1869 in memory of Edward Jones, Bathafarn Farm, 1778–1837 known in the Vale of Clwyd as one of the founding father of Methodism.

Bathafarn Chapel Market Street Ruthin built 1869.
The listed gates and railings by John Williams.

Credit: Gwynfor

Pig Troughs

The pig of these times was a far larger animal and produced fatty bacon that "stuck to the chest in cold weather". Williams made no less than forty six different pig and hog troughs ranging in price from £0 12s 6d to £1 10s 6d. each.

PIG AND HOG TROUGHS.

STRONG PATTERN SINGLE NORFOLK PIG TROUGHS.

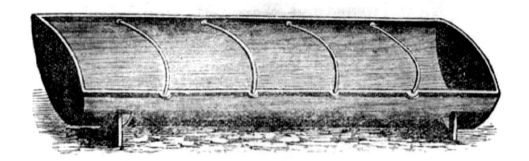

It is intended to stand against a wall or boarding, and so constructed that pigs may feed from it with ease and without wasting the food, or putting their feet into the trough,

No. 1.—2ft.	6in. long, by 8in. wide, for 3 pigs	...
2.—3ft.	0in. do. 8in. do. 4 pigs	...
3.—3ft.	6in. do. 8in. do. 5 pigs	...
4.—3ft.	10in. do. 8in. do. 6 pigs	...
5.—4ft.	8in. do. 8in. do. 7 pigs	...

STRONG PATTERN SINGLE NORFOLK PIG OR HOG TROUGHS.

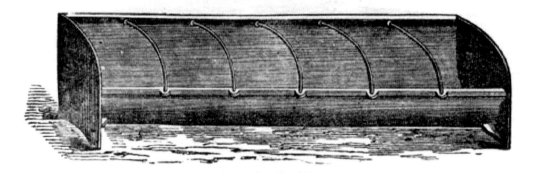

No. 6.—2ft.	4in. long by 11in. wide, for 2 pigs	...
7. 3ft.	1in. do. 11in. do. 2 pigs	...
8.—3ft.	6in. do. 11in. do. 3 pigs	...
9.—3ft.	10in. do. 11in. do. 4 pigs	...
10.—4ft.	6in. do. 11in. do. 5 pigs	...
11.—5ft.	0in. do. 11in. do. 5 pigs	...
12. 6ft.	0in. do. 11in. do. 6 pigs	...

LIGHT PATTERN SINGLE NORFOLK PIG TROUGHS.

No. 13.—2ft. 0in long by 11in. wide, for 2 pigs............
 14.—2ft. 6in.   do.   11in.   do.   2 pigs............
 15.—3ft. 0in.   do.   11in.   do.   3 pigs............
 16.—3ft. 6in.   do.   11in.   do.   4 pigs............
 17.—4ft. 0in.   do.   11in.   do.   5 pigs............
 18.—5ft. 0in.   do.   11in,   do.   5 pigs............
 19.—6ft. 0in.   do.   11in.   do.   6 pigs............

LIGHT PATTERN DOUBLE NORFOLK PIG TROUGHS.

FOR YOUNG PIGS.

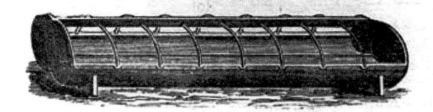

No. 20.—2ft. 0in. long by 12in. wide, for 8 pigs............
 21.—2ft. 6in.   do.   12in.   do.   10 pigs............
 22.—3ft. 0in.   do.   12in.   do.   12 pigs............
 23.—3ft. 0in.   do.   12in.   do.   14 pigs............
 24.—4ft. 0in.   do.   12in.   do.   16 pigs............

More or less divisions can be arranged to the Single Troughs.

STRONG PATTERN DOUBLE NORFOLK PIG TROUGHS.

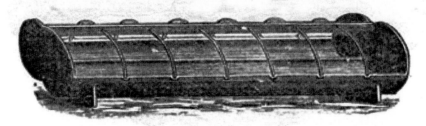

No. 25.—2ft. 0in. long by 16in. wide, for 6 pigs............
 26.—3ft. 0in.   do.   16in.   do.   8 pigs............
 27.—3ft. 10in.   do.   16in.   do.   10 pigs............
 28.—4ft. 6in.   do.   16in.   do.   12 pigs............
 28.—5ft. 3in.   do.   16in.   do.   14 pigs............

HOG TROUGHS.

TWELVE INCHES WIDE, WITH CIRCULAR BOTTOMS AND WROUGHT-IRON CROSS BARS.

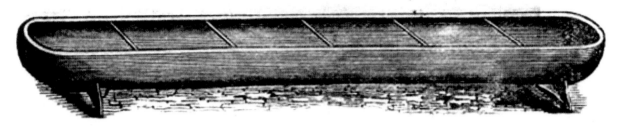

No. 30.—2 feet long ... ...
31.—3 feet long ... ...
32.—4 feet long ... ...
33.—5 feet long ... ...
34.—6 feet long ... ...
35.—7 feet long ... ...
36.—8 feet long ... ...
37.—9 feet long ... ...
38.—10 feet long ... ...

PIG TROUGHS, 10 inches wide, with Circular Bottoms.

No. 39.—2 feet long ... ...
40.—3 feet long ... ...
41.—4 feet long ... ...
42.—5 feet long ... ...
43.—6 feet long ... ...

CIRCULAR IRON PIG TROUGHS.

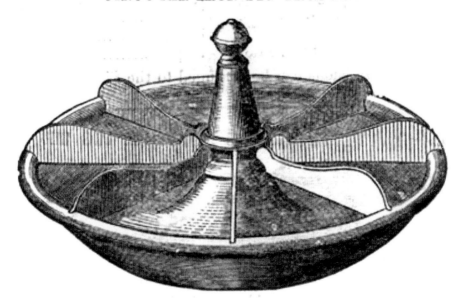

This engraving represents a Circular Trough, with a loose revolving top.

No. 1.—2ft. 0in. diameter, with eight divisions ... ...
2.—2ft. 6in. diameter, with eight divisions ... ...
3.—3ft. 0in. diameter, with eight divisions ... ...

Mangling and Wringing Machines 1875

Growing affluence meant larger houses with a laundry room and the must have machine for the maid was the mangle. These were made in three sizes priced from 67s 6d to 85s. each.

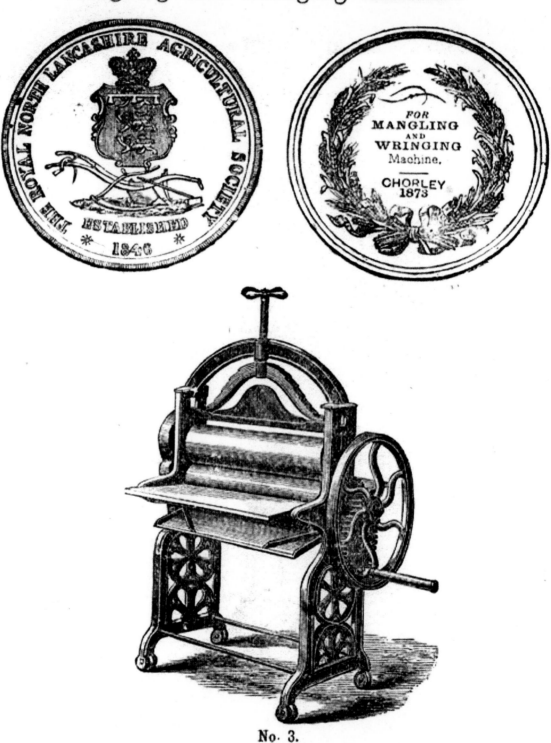

No. 3.

J. WILLIAMS has much pleasure in introducing his improved and newly designed Mangling and Wringing Machine. This is a strong well finished machine, particular attention paid to the rollers being well seasoned.

Price }	24in.	27in.	30in,
	67/6	75/-	85/-

Lawn Mowers 1875

With the building of larger houses and hotels came the interest in lawns, thus the demand for lawn mowers. "The Exceller" Lawn Mowing Machine Made in eight sizes with a price ranging from £3 to £8 each. This machine was exhibited by Williams on Stand No 61 at the 1872 Cardiff Royal Show.

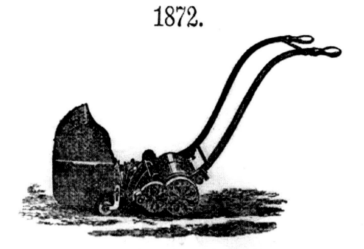

THE "EXCELLER"
Lawn Mowing Machine.
1872.

J. WILLIAMS, in introducing his New Lawn Mowing Machine, begs to say that although there is nothing new or novel in the principle of the machine, he has succeeded in bringing one out that will give universal satisfaction in the working: all that have already been sent out are highly approved of. Every care will be taken in finishing the same in the best manner and out of the best materials, so as to ensure for the machine a large and increasing sale.

Motion is communicated to the knives by machine-made wheel gearing, being far superior and less liable to get out of order or wearing out than many other contrivances which are being resorted to.

The main shaft is steel and the bearings are of hardened steel, so they wear for a very long time without requiring any repairs. The cutting barrels are fitted with the most improved knives. These machines work very silent, will cut longer grass than many others, and are light in draught.

All the grass is collected into a box, which is easily removed when full, or if more desirable the cut grass may be left on the lawn by simply removing the box in front of the machine.

Packing Cases are charged at lowest prices, say, 10, 11 and 12 inches, 3/-; 14 and 16 inches, 4/-; 18, 20 and 22 inches, 5/-; and purchasers are recommended to keep the cases to stow the machines away when not in use in order to protect them from being damaged, or if returned, free of carriage, within one month full price will be allowed for them.

PRICES:—

To cut 10 inches......	To cut 16 inches......
,, do. 11 do.	,, do. 18 do. ......
,, do. 12 do.	,, do. 20 do. ......
,, do. 14 do.	,, do. 22 do. ......

The above Prices include Free Delivery to all the principal railway stations in England.

Improved Corn Drills

Prices as per 1873

	£.	s.	d.
9-Row Corn Drills	21	0	0
10-Row	22	0	0
11-Row	23	10	0
12- Row	25	0	0
13- Row	26	10	0

If with wrought-iron instead of wood levers 3/- each lever extra.

Background

A drill of seed is seed sown in rows. Jethro Tull (a lawyer) was concerned with getting seeds into rows, so that they could be hoed (cleared of weeds) by a horse. His ideas were not fully appreciated until after his death in 1874.

By 1800 the practice of drilling seed on the most productive soils had become universal. James Smyth of Peasenhall Suffolk sent out their machines and offered to do contract drilling at 2s 6d an acre. Smyth became one of the largest manufacturers of grain drills. Founded in 1745, L.R. Knapp of Clanfield Oxon with their "Monarch" drills also established a great reputation.

Faced by such formidable competition by 1875 John Williams stopped making corn drills to concentrate on their turnip and mangold drills.

Cup feed grain drill by Smyth and Sons of Peasenhall Suffolk

Part 1

John Williams

1858-1908

Chapter 3

BARN MACHINERY

(FEED PREPARATION)

Feed

In Winter with the animals housed indoors preparing feed in the barn and feeding could take several hours each day. It was not only farms that housed vast numbers of animals that needed feed. In the towns there were thousands of cows providing fresh milk. In the year 1900 the city Liverpool had 900 registered cow houses (shippons). London had over 10,000 registered cabs with over 50,000 horses involved in transport. Underground haulage in the coal mines was the domain of the pit pony with over seventy thousand in service. The British Army had about 500,000 horses serving both at home and abroad

Thus, feed preparation machinery the mill, chaff cutter, root pulper and cake breaker (barn machinery) became essential equipment for all stock feeders.

By 1850 the feeding of farm animals was based on scientific evidence. The composition and quantities of feeding-stuffs had been calculated and published in journals. A variety of imported concentrates such as cotton and linseed cake were available. A working horse would need about 112lbs of feed daily consisting of mainly milled oats and chopped hay or straw.

Housed indoors in winter a bullock would need daily 3lbs cake, 3lbs meal (milled oats) 80lbs roots, 14lbs of chaffed straw plus some hay.

The Great Exhibition held in London in 1851 was the brainchild of Prince Albert and showcased the best of British Agricultural and Industrial Machinery. I often wonder if John Williams went and that this fired his ambition and his future destiny as a "Maker of Agricultural Implements and Machines".

In 1858 the Royal Show was held at Chester and if Williams had attended he would no doubt be impressed by the advancement made in machinery for the farmer.
This chapter tells the reader how John Williams changed from makers of pig troughs lawn mowers and mangles to becoming one of Britain's premier "feed preparation machinery specialists".

Chaff Cutters

The Horse and the Chaff Cutter

By the middle of the nineteenth century the feeding of the working horse was well researched. Compared with a cow (ruminant-chews its cud), a horse with one stomach is a relatively poor converter of energy. The British Army conducted exhaustive trials and concluded that incorporating chopped straw (chaff) in the feed increased its digestibility and value. With probably close to 3 million horses in the country the market for chaff cutters was enormous.

John Williams became synonymous with chaff cutters thanks to the genius of John Whitaker. In the early 1870s and for the next fifty years Rhuddlan probably made Britain's finest chaff cutting machines.

At the Royal Show held in Cardiff in 1872 Williams exhibited no less than fourteen chaff cutting machines, all invented and improved
by John Whitaker.

Today when we buy a car probably the basic decision is size or model the second is what level from basic to with all the bells and whistles. Williams made four sizes of machines with three level options, thus they were able to provide the customer with a machine that met his exact needs. For example the No 0 machine is a very small hand powered machine, what one may term a "Crofters" model for feeding a few animals.

Prices 1875
No 0 (smallest machine) £2 12s 6d.
No 12 (largest machine) £15 0s 0d.

Our Chaff Cutter

It was about 1970 when I last used the John Williams Son chaff cutter on our farm. I think father took over the machine as part of the farm in 1938. It was originally powered by an Edwards of Llanuwchllyn endless cable drive from a water turbine in a ravine at a distance of a quarter mile!

The chaff cutter built about 1900 never broke down a testimony to both John Williams and my father Thomas Williams who insisted that it was oiled every time it was used. Unlike pulped turnips and swedes, chopped straw did not "go off" so the job was usually done every other day and, of course, on Saturday afternoon as I was available to lend a hand.

Straw chopping (chaff cutting) was the last of the three barn operations. The grain had been milled the turnips pulped and about fifteen sheaves of straw thrown from the loft on to the floor. The feeding trough was attached to the machine and a sheaf laid in the trough and its (band) string cut. Another sheaf would be laid with its head joining the butt of the previous one, with the right hand pressing down on the sheaf about a foot from the feed rollers. With the left hand the feed rollers lever is engaged and the "chaff chaff" sound meant that the cutting action had started. The art now was to feed the sheaves evenly without getting your fingers too near to the rollers. You listened to the sound of the machine, feeding too heavily meant that the top roller moved upwards (rising mouthpiece) lifting the weight off the floor. Our machine would easily cut five sheaves a minute. My job was to provide my father with a steady supply of sheaves and with a shovel clear the machine of chopped straw. The butt of the last sheaf was left between the rollers.

Chopping Bales for Bedding

From the early 1950s the threshing contractors replaced their straw binding machines with balers. The chaff cutter was designed to cut sheaves or loose straw not the compacted twisted straw of a bale. However, the sturdy chaff cutter managed the job admirably without a hitch.

JOHN WILLIAMS,

PHŒNIX IRON WORKS,

RHUDDLAN

PRICE LIST No. 10.

Chaff-cutters.

ALL THESE MACHINES ARE MADE OUT OF THE BEST MATERIALS; THEY ARE
WELL FITTED, AND EVERY CARE IS TAKEN TO TURN THEM OUT
IN THE MOST SATISFACTORY MANNER.

No. O Chaff-cutter, on cast iron legs and frame, toothed rollers
and fast mouth-piece.   Mouth 8 inches by $2\frac{1}{2}$ inches.
Price ....

No. O 1 Chaff-cutter same as No. O, but with large flywheel ...

No. O A Chaff-cutter is the same as No. O, but with a treadle to
be worked by the foot.   Mouth 8 inches by $2\frac{1}{2}$ inches.
Price ....

No. 1—Chaff-cutter suitable for small farmers. The main shaft works in brass bearings. All the wheels are encased in a neat cover. It cuts two lengths of chaff without change wheels, has a rising mouth. Mouth $8\frac{3}{4}$ inches, rising from $2\frac{1}{4}$ to 4 inches. Price.........

No. 1 A—Chaff-cutter is the same as No. 1, but with a larger fly-wheel. Price.........

No. 2—Chaff-cutter is mounted on a cast frame, with large fly-wheel. The main shaft working in brass bearings; has toothed rollers; case-hardened mouth-piece. Mouth $9\frac{1}{2}$ inches wide, rising from $2\frac{1}{2}$ to $4\frac{1}{2}$ inches. Price ....

No. 3—Chaff Cutter is similar to the No. 2, but cuts two lengths of chaff without change wheels. Mouth $9\frac{1}{2}$ inches wide, rising from $2\frac{1}{2}$ to $4\frac{1}{2}$ inches. Price.........

No. 3 A—Chaff-cutter is the same as No. 3, but has a lever for altering the length of cut without stopping the machine.   Price.........

No. 4—Chaff-cutter, is similar to No. 3, but has an auxiliary shaft to help in working the same.   Mouth 9½ inches wide, rising from 2½ to 4½ inches.
                                                                    Price ....

No. 5—Chaff-cutter, is similar to No. 3, but larger and stronger; it has also a stop motion, thereby making it a very safe machine when working with power.   Mouth 9½ inches wide, rising from 2½ to 4½ inches.   Price...

No. 6—Chaff-cutter is similar to No. 5, but has a stop and reverse motion. Mouth 9½ inches wide, rising from 2½ to 4½ inches.   Price.........

PULLEYS EXTRA.

No. 7—Chaff-cutter, for hand or power; it is mounted on a very
    strong cast iron frame, main shaft works in brass bearings,
    has a case-hardened mouth-piece, a heavy fly wheel with

two knives, toothed rollers, with stop and reverse motion, cuts two lengths of chaff without change wheels, and has a rising mouth-piece. Mouth $10\frac{1}{2}$ inches wide, rising from $2\frac{1}{2}$ to $5\frac{1}{2}$ inches. Price ....

No. 8—Chaff-cutter is the same as No. 7, except that it is provided with a clutch gear, which, by moving a lever, enables the person feeding the machine to alter from one length of cut to another without stopping the same. Mouth $10\frac{1}{2}$ inches, rising from $2\frac{1}{2}$ to $5\frac{1}{2}$ inches. Price ....

No. 9—Chaff-cutter, is similar to No. 7, but larger and stronger. Mouth 12 inches wide, rising from $2\frac{1}{4}$ to $5\frac{1}{2}$ inches. Price

No. 10—Chaff-cutter is the same as No. 7, but with clutch gear to alter the length of cut without stopping the machine. Mouth 12 inches wide, rising from $2\frac{1}{2}$ to $5\frac{1}{4}$ inches. Price

No. 11—Chaff-cutter is similar to No. 9, but much larger, and stronger. Mouth $13\frac{1}{2}$ inches wide, rising from $2\frac{1}{2}$ to $6\frac{1}{2}$ inches. Price ....

No. 12—Chaff-cutter is the same as No. 11, except that it is provided with the clutch gear and lever to alter the length of cut, without stopping the machine. Price...................

PULLEYS EXTRA.

Length of Chop

By 1850 feeding of farm animals had becoming a science. Chopped straw (especially oat straw) formed an important part of diet. The optimum length of chopped straw for a ruminant animal (cow) should be 3/4inch and for a horse (non ruminant) 3/8inch.

On our farm we chopped at 3/8inch. Some chopped straw was also used for bedding down and speeded up the rotting of the resulting manure.

The 1897 Chaff-Cutting Machines Accidents Act

The Act stated that "All powered chaff-cutters to have an apparatus to prevent the hand or arm being drawn between the feed rollers into the knives. The device to be of modest cost and easily adapted for fitting to existing machines."

Trials of Methods of Safe-guarding Chaff Cutters to comply with the Act of 1897 were held at Four Oaks Park Birmingham Friday the17th of June 1898. There is no evidence of Williams entering this trial but from the first of August 1898 their powered machines would have to comply with the Act.

The judge of the trials was Mr R.M Greaves Wern Portmadoc, North Wales.

By this time John Whitaker had left John Williams to form Powell Brothers and Whitaker at Wrexham, they entered the competition.

Most manufacturers came up with the idea of a crossbar when pushed reversed the drive of the feed rollers. Mr Greaves was critical of this method reasoning. "In the event of the attendant being drawn into the feed rollers it is doubtful whether he will have the presence of mind to bring his body forward to contact the cross bar, the tendency would be to endeavour to withdraw the hands by throwing the body backwards thus delaying the action of the cross bar until too late".

Author's experience "I never fed the rollers with both hands in the forward position".

*Chaff- cutter gear change, giving neutral, two lengths of chop and safety reverse
(patented by John Whitaker).*

Credit: Gwynfor

Conclusion

The fate of the chaff cutter is linked to that of the horse and during the 1950s thousands of chaff cutters were scrapped for a few shillings. At the time of writing it is estimated that there are a million horses and ponies in the UK. The days of the "old chaff cutter" are not quite numbered yet.

Machines in preservation

Of the thousands made at Rhuddlan, I estimate that about forty still exist, some in museums, private collections. There are still a few in barns or lying neglected in the open. There is a John Williams & Son machine at the Greenfield Valley Museum and Heritage Park Holywell.

The Root Pulper

two knives, toothed rollers, with stop and reverse motion, cuts two lengths of chaff without change wheels, and has a rising mouth-piece. Mouth $10\frac{1}{2}$ inches wide, rising from $2\frac{1}{2}$ to $5\frac{1}{2}$ inches. Price ....

No. 8—Chaff-cutter is the same as No. 7, except that it is provided with a clutch gear, which, by moving a lever, enables the person feeding the machine to alter from one length of cut to another without stopping the same. Mouth $10\frac{1}{2}$ inches, rising from $2\frac{1}{2}$ to $5\frac{1}{2}$ inches. Price ....

No. 9—Chaff-cutter, is similar to No. 7, but larger and stronger. Mouth 12 inches wide, rising from $2\frac{1}{2}$ to $5\frac{1}{2}$ inches. Price

No. 10—Chaff-cutter is the same as No. 7, but with clutch gear to alter the length of cut without stopping the machine. Mouth 12 inches wide, rising from $2\frac{1}{2}$ to $5\frac{1}{2}$ inches. Price

No. 11—Chaff-cutter is similar to No. 9, but much larger, and stronger. Mouth $13\frac{1}{2}$ inches wide, rising from $2\frac{1}{2}$ to $6\frac{1}{2}$ inches. Price ....

No. 12—Chaff-cutter is the same as No. 11, except that it is provided with the clutch gear and lever to alter the length of cut, without stopping the machine. Price..................

PULLEYS EXTRA.

Improved Disc Root Pulper and Slicer.

These machines consist of a strong cast iron disc with a heavy rim, acting as fly wheel, mounted on an iron frame, with wood legs, and an improved hopper, made so that the last piece must be cut. They are very easy to be worked. They are made with two knives for Slicing Turnips, and also as a Pulper. The Pulping Knives can be moved so as to pulp fine or coarse, by simply slacking the nuts and setting the same more or less forward. The Disc and Knives are covered with an Ornamental casing.

Price as a Slicer, No. 0, with Two Knives ....
  ,,   ,,   No. 1, do. do.   do.   ....
  ,,   ,.   No. 2, do. do.   do.   ....
Price as a Pulper, No. 3, with Six Knives ....
  ,,   ,,   No. 4, ditto Eight ditto ....
  ,,   ,,   No. 5, ditto Ten ditto   ....

PULLEYS EXTRA.

NEW IMPROVED

Combined Machines for Pulping and Slicing.

These machines are fitted with two cast iron disc wheels, the one has suitable knives for pulping, which are so arranged as to enable the man feeding the machine to alter the cut to fine or coarse pulp, as may be required. The other disc is fitted with two knives of an improved shape for slicing : they are al-tered from a pulper to a slicer by merely reversing a lid inside the hopper. They require much less power to work them than the generality of machines used for the above purposes.

No. 6—Double-action Pulper and Slicer ....
  ,, 7—   Ditto   ditto   ditto   ....
  ,, 8—   Ditto   ditto   ditto   ....

Stripper Knives for No. 6 and 7, 10/- extra; ditto, ditto, for No. 8, 11/- extra.
PULLEYS EXTRA

In the 1950s on our mixed farm of 47 acres near Ruthin we grew 3 acres (about 50–60 tons) of roots. What a job on a cold late October morning. Each root had to be pulled out and tailed and topped and put in rows loaded by hand on a cart or trailer and taken to the farm.

There they were unloaded by hand and thrown through a shutter type door on to an earth floor store in the barn. When the store was full they were piled up against the barn wall and thatched with straw to protect from frost. A nearby hedge afforded protection from the sun.

Adjacent to the store were the hand and engine powered root pulping machines.

According to a census in1889 there were over two million acres of roots grown in Great Britain. The county of Denbighshire boasted the largest acreage in the whole of Wales. Turnips swedes and mangolds were once the mainstay of the cattle's winter diet. Patented in 1839, Gardner's turnip Slicer was the premier machine. Samuelson's of Banbury had made thousands. When turned one way the machine produced the preferred finger-shaped pieces and when reversed made slices (double-action).

Such was the success of this machine that many of the leading manufacturers took up the rights to manufacture this cylindrical turnip pulping machine in both single and double action.

Amongst these were John Williams & Son who made this machine from about 1870 to 1920.

Disc Root Slicers and Pulpers

Whilst at Picksley and Sims John Whitaker had filed patents with reference to Root Pulpers dated November 10th 1862 Pat No3026 and June 14th 1866 Pat No1621.

At the 1872 Royal Show held in Cardiff on Stand No 61 Williams exhibited nine machines ranging in price from £3 5s to £7 10s. all invented by John Whitaker and manufactured by the Exhibitor. It is interesting to note that all these machines were fitted with a handle and that a pulley for horse, water power or steam was extra.

Machines in Preservation

At the time of writing only four are known to exist. During the 1950s the practice of pulping mangolds, swedes and turnips declined. Machines were scrapped in their thousands.

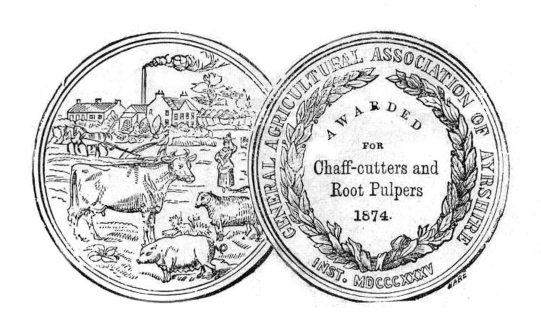

Feed Mills

Oilcake Mills (Oilcake Breakers)

Today cattle cake is mainly fed in the form of cubes or pellets. It comes in a wide range of formulations to meet the animal's precise dietary needs. It is delivered in bulk in loads of twenty tonnes or more.

For many years up to the 1960s it was delivered by rail in hessian bags (later paper sacks). From the railway station it was delivered by lorry to a depot or on to the farm.

A small farmer would on market day pick up a few bags in the boot of his car.

Cake is the by-product obtained after oil is extracted from a variety of seeds such as cotton seed, palm kernel and linseed.

As soap manufacturers, Lever Brothers were importing vast tonnages of palm kernel and producing the valuable by-product of cake. It was therefore almost by accident that they became amongst the foremost of cattle feed millers.

Bibbys who were millers in Lancaster moved to Liverpool to expand their business.

Criddle started importing tea coffee and molasses (syrup and treacle).

Cake was made into blocks using molasses as a binder and so had to be broken up for feeding the animal.

In the 1875 Catalogue the John Williams Single Action Oil Cake Mill; invented by John Whitaker was priced £3 10s. The Double Action Oil Cake Breaker with extra pair of rollers which can be used to make a finer sample for lambs and sheep was priced £7. These machines remained in production for almost fifty years.

No. 1 - £3 10s   *No. 2 - £7*

Grist Mills

In a rural area the most powerful machine was the water mill or a farm that was fortunate to have a water wheel.

On a farm one of the duties of "the carter" was to take the grain to the mill to be ground for both domestic and animal feed.
Up to about 1850 there were no mills suitable for farm use and, if they were, the power to work them was also not available. The development of the chilled cast iron plate grist mill coupled with an oil engine gave the farmer control of milling for animal feed.

Grain milling for human consumption had been taken over by the larger milling companies who built large steam powered dockside mills at ports such as Liverpool and Hull.

John Williams & Son Grist Mills

As one of the leading manufacturers of (barn) feed preparation machinery the company must have realized that to make their range complete they needed to add a grist mill to their portfolio.

The 1890 RASE Trials of Grist Mills held at Plymouth set out to find the best grist mill for farm use with an engine not exceeding 10 horsepower. John Williams and the leading manufacturers of the day entered the trials which took place over two days in June.

Each competitor was given 56 lbs of various seeds to be ground to a predetermined degree of fineness. The time and power required would be recorded. The first prize of £20 was awarded to S. Corbett of Wellington Salop. Their machine, amongst the cheapest at £13, ground 56 lbs of oats in less than three minutes consuming an average 3.97 horsepower.

As the Williams mill priced at £15 did not feature in the final listings there are no data as to its performance. The trial proved the supremacy of the chilled cast iron plate mill.

Author's Note

On our farm we had a Bamford's No2 mill which ground 112lbs (1cwt) of oats in about twenty minutes. As the plates became worn output was reduced and the meal became warm reducing its keeping quality. To restore performance the plates are turned over (reversed) to reveal a new working surface.

Machines in preservation

A few (about three) cake breakers are known. The author at the time of writing does not know if any Williams Grist Mills survived. Examples of John Williams implements and machines can be seen at the Greenfield Valley Museum and heritage park Holywell North Wales.

Horse Power Works
(also known as horse gears or horse gin)

A means of converting animal power to rotary motion. The invention of the threshing machine and barn machinery meant that the farmer needed power. Williams responded by manufacturing two models for one and two horses.

Butter churning and root pulping needed a slow output speed; chaff cutting and threshing a higher speed. This higher gearing would be met at extra cost by an intermediate motion.

From about 1850, picture the scene on countless farms the horse or horses harnessed to the "power" threshing, root pulping and chaff cutting.

Note I have not included grain milling as this would have been done by the local watermill

My mother, on the farm of Maestyddyn Isa Clawddnewydd near Ruthin during the 1920s, used to recall butter churning with a pony gear. The wall of the dairy was built into the earth slope to keep it cool and through a shutter window at eye level she could see the hooves of "Nedw" harnessed to the power. She would say "m'laen Nedw" and the pony would proceed turning the gear. To stop churning the command "we Nedw" was given.

By 1930 most horse gears had fallen out of use. Horse power had given way to oil power. See Oil Engine.

Improved Horse Power Works.

This Horse-power Works is very compact, the gearing being covered in by a neat cast iron dome : is very easy to work, and requires but little expense to fix.

PRICE :

One Horse Power Works   ...   .
Intermediate motion for same, with one 17 inch Pulley   ...   ...   —   ...
Two Horse Power Works   ...   .
Intermediate motion for same, with one 17 inch and one 14 inch Pulley...   ...   ...

New Improved Horse Power Works.

This Horse Power Works has three different speeds for churning, chaff-cutting or pumping. The intermediate wheels are enclosed by a neat cover, which is not shewn.

Price (complete) £10 10   0

Machines in preservation

Four horse power works manufactured by Williams are known to exist.

Part 1

John Williams

1858-1908

Chapter 4

FIELD MACHINERY

ROLLERS HARROWS AND HOES

Flat Field Roller

The field roller has three primary uses:

• To crush clods

• To consolidate the soil (firm seed bed)

• To push stones back into the turf on grassland, thus eliminating damage to the harvesting machinery (mowers).

The 1874 price list notes Williams manufacturing no less twenty four different models of roller in three main types. This vast range is achieved by a combination of ring (segment) diameters and numbers of rings = width. The smallest and would be 5 feet wide up to the largest at 8 feet.

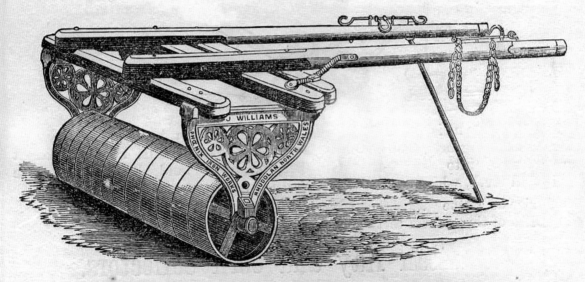

Improved New Pattern Field Roller.

This implement is composed of a series of rings with flat surfaces, which, working independently on a round spindle, adjust themselves to the undulating surfaces much better than the ordinary barrel roll, and is more easily turned at the ends of fields. It is also so arranged as to reduce all friction from the axle to the smallest possible extent, being fitted with Grease or Oil Boxes. The above may be had fitted with patent tubular iron frames and shafts at the same prices.

Width. ft. in.	14-in. dia. £ s. d.	16-in. dia. £ s. d.	18-in. dia. £ s. d.	20-in. dia. £ s. d.	24-in. dia. £ s. d.

Improved Cambridge Pattern Clod Crusher

J. WILLIAMS'S
Improved Cambridge Pattern Clod Crusher.

This Clod Crusher having been in use for some time now, continues to be very highly spoken of. It is composed of a number of wheels with thin cutting edges, 3 inches wide, each turning freely on the spindle. It will cut through and break the hardest clods and reduce the small ones to dust. It is extensively used for meadows and Pasture Lands, for Parks and Lawns, the hard surface being broken, the grass soon becomes thick and luxuriant. It is effectual for re-setting Wheat, Vetches and other Plants made light by frosts.

It Stops the Ravages of Wire Worm.

The Corn soon tillers out and covers the ground that previously appeared like a piece of Fallow. It surpasses all Flat Rolls for the following purposes :—

For Rolling Cloddy Land before Harrowing ; also for Rolling Corn as soon as Sown.
For Rolling Wheats upon Light Land in the Spring, after frosts and winds have left the Plants bare.
For Crushing Clods after the Turnip Crops. to sow Barley.
For Stopping the Ravages of Wire Worm and Grub,
For Rolling Barley, Oats. &c., when the Plants are three inches out of the Ground.
For Rolling before Sowing Clover again in Autumn, and in Winter or Spring whenever the Clover Plant has a tendency to throw out.
For Rolling Turnips in the Rough Leaf before Hoeing, when the Plants are attacked by the Wire Worm and Grub.
For Rolling Grass and Mossy Lands after Compost.

Width.	16in. dia.	18in. dia.	20in. dia.	22in. dia.	24in. dia	26in. dia.
ft. in.	£ s. d.	£ s. d.	£ s. d.	£ s. d	£ s. d.	£. s. d

N.B.—J. W. can Supply Clod Crushers with SEGMENTS of any width ; also RINGS, SPINDLES, and BRACKETS separate, so that parties having Timber may make up their own Rollers.

The New Patent Spencer's
Zig-Zag Roller and Clod Crusher

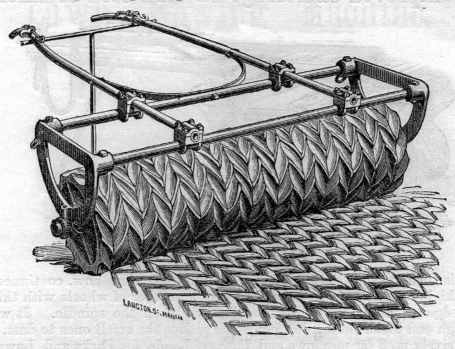

THE NEW PATENT
Cambridge Zig-Zag Roller and Clod Crusher.

LANGTON, S.T. MANCR

The New Patent Roller and Clod Crusher will surpass every other in use for the following purposes :—For Rolling Cloddy Land before Harrowing ; also for Rolling Corn as soon as sown, For Rolling Wheats upon light land in the Spring, after frost and winds have left the plants bare. For Crushing Clods after the Turnip Crops, to sow Barley. For stopping the ravages of the Wire Worm and Grub. For Rolling Barley, Oats, &c., when the plants are three inches out of the ground. For rolling before sowing Clover again in Autumn, and in Winter or Spring whenever the Clover plant has a tendency to throw out. For Rolling Turnips in the Rough Leaf before hoeing, when the plants are attacked by the Wire Worm and Grub. For Rolling Grass and Mossy Lands after compost.

Some of the peculiar advantages of this Roller over others are, the equalization of the pressure upon the land. The discs, which work independently of each other, do not sink so deep into the land ; therefore the draught is lighter. The indentions of the discs approach nearer to the tramping of sheep than any other, and a special advantage is found in the Rolling of Wheat and other crops in Spring—in the destruction of the Wire Worm, Grub, &c.; it also obviates the objection to the Clover Seed coming immediately in the row of Corn.

PRICES AND DIAMETER OF WHEELS.
WITH IRON FRAME AND TUBULAR SHAFTS.

Wheels.	20 in. diameter.	24 in, diameter.	26 in diameter
Five feet wide			
Six feet wide			
Seven feet wide			
Eight feet wide			

If with Wood frame and shafts £1 each size less, if with double Shafts £1 5s. 0d. extra.

Horse Hoes

Three-tine Horse Hoe.

The above is most useful for Hoeing one row of Turnips, Mangold Wurzel, Beans, Potatoes, &c. It is made entirely of Wrought Iron, with the exception of the wheel which, is Cast Iron. The teeth can be moved to any width required.

Price of 3-tine Hoe, Light...   ...   .
Ditto of do.   do. Heavy   ...   .

The Three-tine Horse Hoe. Two versions available, light and heavy.

Five-tine Horse Hoe and Grubber.

This Implement is much improved, and in addition to its efficiency as a Hoe for cleaning between Turnips, Mangold Wurzel, Beans, Potatoes, &c., it can be made into a Five-tine Grubber, on light land, by having three additional teeth, which can be substituted for the front and hind ones.

Price as a Hoe ...   . £
Ditto of three extra teeth when required   ...   ...

The Five Tine Horse Hoe and Grubber

Machines in Preservation

Under normal use a roller has a long life. Horse drawn rollers were converted for use with the tractor and may have been in use as late as the early 1970s.
Many were scrapped but a few have survived in a barn or hedgerow.

Seed Sowing

Root Drills

At the Cardiff Royal Show of 1872, John Williams exhibited two machines:
Exhibit No1350 Single Row Turnip and Mangold Drill Price £3. The quantity of seed sown is regulated by three different sized pulleys worked by a strap.

One-row Turnip and Mangold Drill.

This is a very excellent Drill similar in principle to the Two-Row Drill.

Price ... ... £

Exhibit No1351 Double Row Turnip and Mangold Drill. Price £6 6s, two sets of ladles (cups) and change wheels to regulate the quantity of seed to be sown.

Two-row Turnip and Mangold Drill.

This is a most excellent Drill for Turnips, Mangolds, &c., with Cup Ladles to raise the Seed, and Change Wheels to regulate the quantity to be sown.

Price ... ... £

These two machines were to remain in production for almost fifty years and worked on the cup feed system.

The seed hopper is divided into two parts, an upper compartment, the grain box and the lower the feed box. In the feed box is a disc with a series pf spoons on its periphery.

Each cup picks up a few seeds from the feed box and at the top of its turn, the seeds fall from the spoon into the coulter tube and the ground.

As the seed was multi-germ, two or more plants would emerge from one seed and so thinning and singling was an important part of growing the crop. See Corbett-Williams Turnip Thinner.

The "New Era" Turnip and Mangold Drills single and two row versions.

These were made by several companies starting manufacture in 1895. As many as 800 a year were made. The machine is simplicity itself. Rotating brushes agitate the seed through a calibrated hole.

Credit: Paul Jones

Machines in Preservation
A two row machine at the Greenfield Valley Museum and Heritage Park Holywell.
A single row machine at the National Museum of Wales (Welsh Folk Museum St Ffagans).
There are several machines in private collections.

Mowers and Reapers

JOHN WILLIAMS,

Phœnix Iron Works,

RHUDDLAN, Near RHYL,

NORTH WALES.

PRICE LIST OF

MOWING & REAPING

MACHINES,

For the Season 1875.

These Machines have, in various important Agricultural Centres in the United Kingdom, been awarded

GOLD & SILVER MEDALS & FIRST PRIZES,

In Competition with those of the leading Makers.

By 1860 it could be argued that the horse drawn mower had reached a fair degree of perfection. Amongst the leading companies were Picksley and Sims of Leigh who manufactured the Bamlett Patent Reaping Machine for cutting grain and grasses.
At the Trials of Reaping Machines held at Garforth near Leeds, Picksley and Sims gained first prize.

No doubt the man behind this success was no less than John Whitaker who on moving to John Williams circa1870 brought with him a wealth of experience.

The FIRST PRIZE at the Writhlington Agricultural Society's Trials, held at Wellow, near Bath, on the 5th June, 1874, was awarded to John Williams's "Victoria" Mowing Machine.

SHOWN AS A MOWER.

PRICES:—

As a Mowing Machine only, to cut 4ft. 6in. wide, with two knives, box of tools, and a supply of small wearing parts £22   0   0

As a Combined Mowing and Reaping Machine, complete as above   26   0   0

Best double-stitched Collar Straps, 10/6 per pair.

The Victoria Combined Reaping and Mowing Machine.

This Machine has been well tested as a Mower and Reaper, and has given great satisfaction. It is all made of iron and steel, out of the very best materials that can be procured, so as to make it one of the best and least liable to breakages of any that are in use. It is the least complicated out—all the parts are very easy to get at in case of repair, and great care is taken in the workmanship. The finger bar has steel bottom knives affixed within each finger, thereby enabling the same to be sharpened when required; it thus forms a sharp cutting resistance to the knife similar to a shears, and will cut any crop, however laid, with great ease. Both the driving wheels assist in working the knife. The finger bar is carried lightly over the land on two wheels by which the length of the cut is regulated. The finger bar can be raised and lowered by the driver, with one hand, whilst seated.

SHOWN AS A REAPER.

At the Cardiff Royal Show of 1872 John Williams introduced the Clwydian Two-Horse Mowing Machine exhibited as a (New Implement) priced £20. As a reaper Price £25
The above price includes one extra knife, box of tools, and a supply of small wearing parts such as knife sections and rivets.
At the same show also exhibited as a (New Implement) was a One-Horse Reaping Machine Priced £16 10s.

These two machines probably formed the basis for the "Princess" Light and the "Victoria" Heavy Combined Machines who gained several prizes during 1874. Both machines were

DIAGRAM OF WEARING PARTS OF WILLIAMS'S VICTORIA MOWER AND COMBINED MACHINES.

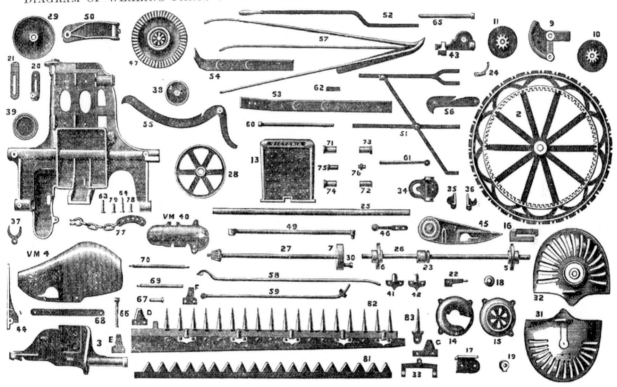

The Princess New Combined Mowing and Reaping Machine.

THE FIRST PRIZE
GOLD MEDAL,
OF THE
CORK FARMERS' CLUB.
AT THEIR
ANNUAL TRIALS,
HELD AT
BLARNEY,
NEAR CORK,
On the 21st & 22nd, July.
1874.

PRIZE MEDAL,
AYRSHIRE
AGRICULTURAL
SOCIETY.
APRIL, 1874.

Price as a Mower with 4 feet bar...	...	...	...	...	...	...	£20	10	0
Ditto Ditto with 4ft. 3in. bar	...	...	...	...	...	...	21	0	0
Ditto as Combined Mower and Reaper	...	...	...	...	...	...	25	0	0

held in high esteem at the Royal Agricultural Society of England Trials of Mowers and Reapers held at Taunton 1875.

Both machines built on the same frame Victoria 2ft 82in diameter wheels width of cut 4ft 6in while the Princess had 2ft 6in diameter wheels width of cut 4ft.
1884 Reaper for one horse "Phoenix"
By 1880, the leading British manufacturers were Bamlett of Thirsk, Harrison Mc Gregor of Leigh (Albion), Bamford's of Uttoxeter. From America poured in Massey and Harris, McCormick, Deering and others.

SHOWN AS A REAPER.

Price of Combined Mowing and Raeping Machine £25   0   0

WEARING PARTS OF WILLIAMS'S PRINCESS NEW COMBINED MOWING AND REAPING MACHINES,

When ordering spare parts give the number on the casting and the number of the Machine.

A new challenge came from Wrexham the Powell Brothers and Whitaker "Cambrian Mowers". Despite being highly regarded, John Williams mowing machines were never sold in large numbers.

Machines in preservation

The Princess Mower asleep for over 100 years in a wood near Pwllglas.
Left Huw, middle, Gareth, right Rhys.

Credit: Gwynfor

Only one known a "Princess" discovered in 2012 on farmland at Pwllglas near Ruthin.

John Williams, Phœnix Works, Rhuddlan, near Rhyl.

WILLIAMS'S

'VICTORIA' AND 'PRINCESS'

MOWING AND REAPING MACHINES,

HAVE BEEN AWARDED THE FOLLOWING PRIZES AND MEDALS,

In Competition with Hornsby & Sons; Samuelson & Co.; W. A. Wood (American); D. M. Osborn & Co. (American); Harrison, McGregor & Co.; Bamlett's, &c. :—

The Silver Medal of the Manchester and Liverpool Agricultural Society, Chester Meeting, 1873.

The Silver Medal of the Royal North Lancashire Agricultural Society, Chorley Meeting, 1873.

Second Prize at the West Gloucester Farmers' Club Trials of Mow-ing Machines, held at Stapleton, near Bristol, June 11th, 1873.

The Prize Medal of the Ayrshire Agricultural Society, Scotland, at their Annual Agricultural Meeting, held at Ayr, April 28th and 29th, 1874.

The First Prize in the Manufacturers' Class of the Writhlington Agricultural Society, at their Annual Trials at Wellow, near Bath, held on the 5th of June, 1874.

The First Prize Gold Medal of the Cork Farmers' Club, at their Annual Trials held at Blarney, near Cork, on the 21st and 22nd July, 1874.

The Silver Medal of the Manchester and Salford Fat Cattle Show, Royal Pomona Gardens, Manchester, Nov. 20th, 21st, 23rd, 24th and 25th, 1874.

Most satisfactory Testimonials can be given of all Machines sent out.

MANUFACTORY:

PHŒNIX IRON WORKS, RHUDDLAN,

NEAR RHYL.

Rakes and Hay Collectors

Hand Drag Rakes
At Cardiff, in 1872 no fewer than nine different hand rakes were exhibited ranging in price from 10s to 15s and 6d.
American Hay and Corn Collectors
Priced at £2 with 15 hickory teeth.
Price £2 5s

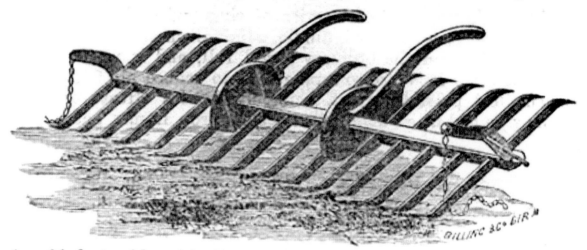

American Hay and Corn Collectors.

No. 1.—9½ feet wide with 12 Ash Teeth..........................
No. 2.—9½ feet do. with 15   do   ....
No. 3.—9½ feet do. with 23   do   ....
If made with Hickory Teeth 3s. each rake extra.

Wood Framed Hand Drag Rakes.
WITH IRON SCREWED TEETH. PAINTED RED.

PRICE :

22	24	26	28	30	Teeth.
4ft. 6in.	4ft. 8in.	5ft.	5ft. 6in.	5ft. 10in.	lgth. of hd. each.

If required for Exportation, 6d. each extra.

Part 1
John Williams
1858-1908

<u>Chapter 5</u>

DARK DAYS

THE JOHN WILLIAMS FUND 1897

In 1895 William Bridge Williams aged 44 took over at Rhuddlan from the now ailing John Williams.

The years between 1874 and 1896 constituted a period of falling prices of animals and farm produce. In 1877 the potato crop failed, and sleet during the summer of 1879 saw fungus disease of the corn crop, with pneumonia affecting cattle and sheep. A succession of poor summers in 1891–92 was followed by drought in 1893.

The late 1880s had proved to be hard years for John Williams & Son. They faced fierce competition from a host of companies amongst them the other Welsh manufacturer of Powell Brothers and Whitaker at Wrexham. Two years later, we read in The Review dated July 1897 that a testimonial fund had been set up on behalf of John Williams. In May of that year at Rhyl Town Hall "The John Williams Fund "was set up. Most of the subscriptions were obtained in the North Wales District and Liverpool.

The officials were as follows: Abel Jones, Gordon House (chairman) Samuel Perks J.P. Dolanog (treasurer) Jacob Jones, Clifton Villa (joint honorary treasurer) and R.W. Jones manager of the African Oil Mills Company Liverpool.

It is interesting to note that subscriptions could be sent to the North and South Wales Bank or the London and Provisional Bank Rhyl. It was reported that by July £217 10s 6d had been collected.

Photo circa 1895. A frail John Williams in top hat enjoying the sea air on Rhyl promenade.
Credit: Mr & Mrs John Dod (Frank Jones)

At the Royal Agricultural Society of England's meeting at Manchester the following agricultural implement manufacturers donated:

H. Bamford & Sons; Barford & Perkins; Barker & England; E.H. Bentall &Co; Blackstone & Co Ltd; T. Bradford &Co; Clayton & Shuttleworth; Thos Corbett;
Christy & Norris; H. Denton; R. Greenough; Harrison Mc Gregor & Co Ltd;
T. Holyoak & Sons; R. Hornsby & Sons Ltd; J & F Howard; R. Hunt & Co;
T. Jones & Co; Kearsley & Co; R.A. Lister & Co Ltd; Mc Hattic & Co; Marshall Sons & Co;
S. Osborn & Co; W. Patton & Sons; Powell Brothers and Whitaker; J.C. Prestwick;
Ransomes Sims & Jefferies Ltd; Samuelson & Co; Tyzack, Sons & Turner; Waide & Sons;
John Wallace & Sons: Woodroofe & Co.

By November 1897 a total of £368 had been received and on the sixteenth of that month at a meeting held at the County Offices at Rhyl the first instalment of £300 was presented to John Williams. It appears that the final total came to over £400.
The people of Rhuddlan among whom he had mixed so much, were also generous and on their own account presented him with a suitable testimonial.

John Williams had served the Phoenix Ironworks up to his eightieth year, he could now be free of anxiety and spend his remaining time as pleasant as possible.

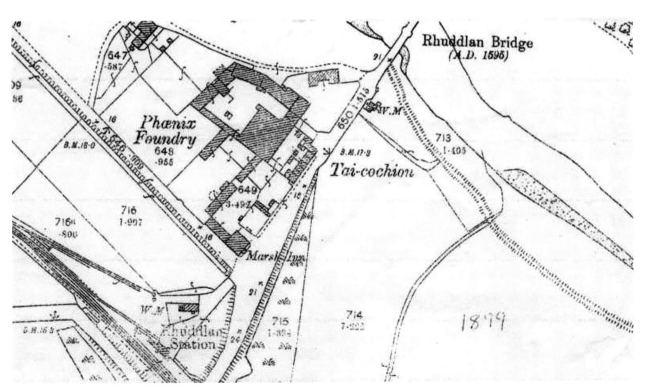

Area occupied by Phoenix Foundry circa 1899.

Death of The Reverend John Williams

On Thursday January 19th 1899 aged John Williams died aged 83 years.
"The funeral will take place on Monday the 23rd with interment in the old Church Cemetery Rhys."

I quote from a copy of the Rhyl Journal dated January 21st 1899.

"We have with much regret, to announce the death of the Rev John Williams, which took place at his residence in Elwy Street, on Thursday morning. The deceased had been gradually sinking for some time past, he having reached the ripe age of 83 years. From a boy he had taken the deepest interest in religious matters. At the Vale of Clwyd Calvanistic Methodist meeting held in Rhuddlan in January 1896 he was presented with an illuminated address by the members. Williams had joined the Calvanistic Methodists aged 16 years whilst living near Betws yn Rhos. Aged 18 he was elected deacon. Aged 25 years he and his wife Eunice took up residence in Rhyl where here he delivered his first sermon. Mr Williams for many years carried on business as an ironmonger and was founder and proprietor of the Rhuddlan Foundry. He retired from the Board of Commissioners for Rhyl in December 1893".

There were few men in the Vale of Clwyd better known than Mr Williams and he had often been referred to as one of "The Fathers of Rhyl".

The Phoenix Ironworks circa 1908. On the side of the building is "John Williams & Son Ironfounders Agricultural and Motor Engineers".

Credit: Mr & Mrs John Dod (Frank Jones)

Part 2

Corbett-Williams & Son

1908-1923

Chapter 6

CHANGE OF

OWNERSHIP

CORBETT-WILLIAMS AND SON

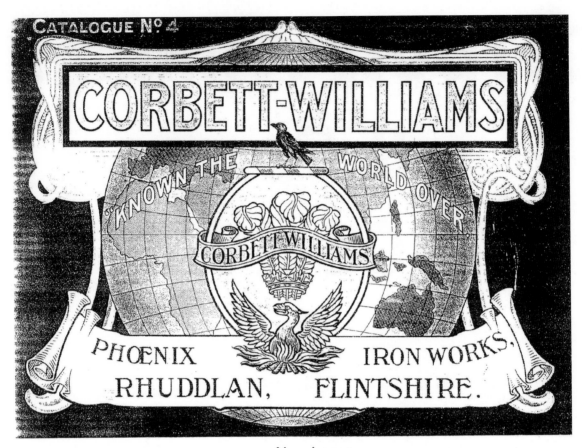

New Image

The Implement and Machinery Review Dated January 1st 1909 stated:

CHANGES OF PROPRIETORSHIP IN A WELSH IMPLEMENT-MAKING CONCERN

In December 1908 Mr Francis Corbett of Wellington Shropshire bought into the Phoenix Iron Works.

Francis Corbett then became chairman and managing director of Corbett, Williams & Son. Francis was the youngest son of the late William Corbett a senior partner in the old-established firm of S. Corbett & Son Park Street Wellington Salop.

At the 1890 RASE "Trials of Grist Mills" from an entry of sixteen, Corbett's secured first prize "for the best grist mill for use on a farm to be worked by an engine of not more than 10 brake-horsepower".

Francis Corbett
Credit: Mr & Mrs John Dod (Frank Jones)

The Corbetts were keen shrewd businessmen and were related to Thomas Corbett of the Preservance Works Shrewsbury.

At Rhuddlan Francis Corbett appointed F.A. Goldsmith as works manager. Goldsmith was a man who had many years service with John Williams & Son and was renowned as a hard working man. He had been apprenticed to the steam engine builders Charles Burrell of Thetford. Following this he was associated with Philip Pierce of Wexford who had gained a reputation for their mowing and reaping machines.

The Goldsmith family lived at 4 Marian Villas, Parliament Street, and Sunnyside, Rhyl Road, Rhuddlan.

Age had thrown its dusky mantle over the Phoenix Iron Works, but things were about to change under the directorship of Francis Corbett.

Frank Arthur Goldsmith
Credit: Brenda Wyatt

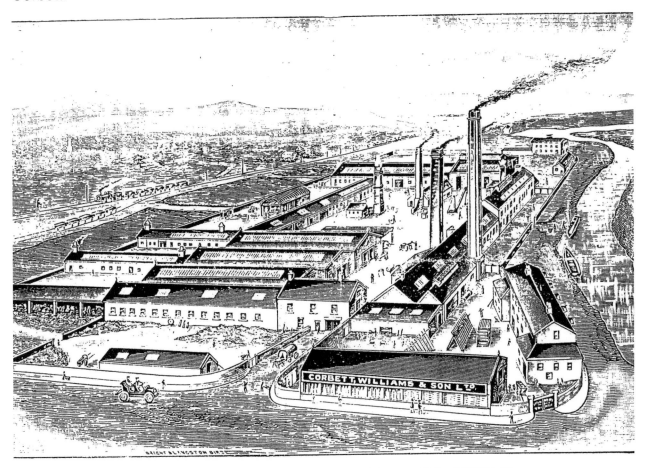

CORBETT, WILLIAMS & SON'S PHŒNIX IRON WORKS, RHUDDLAN.

Artistic licence!

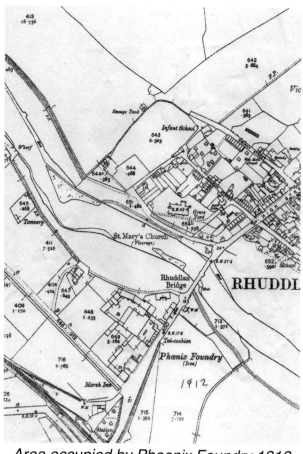

Area occupied by Phoenix Foundry 1912.

His first move was to embark on a "world tour" aimed at increasing sales abroad. Second, he invested in new buildings and machinery. I quote from The Implement and Machinery Review May 1st 1910.

"The extension to their works have been on a large scale. They have erected new paint shop, increased the area of their fitting shop by 120ft by 35ft; built new stores and anew experimental shop. They have installed new Roots blowers and new cupolas and have laid down two 42 h.p. suction gas engines and two suction gas plants manufactured by Fielding and Platt of Gloucester. At the present time they are adding to their workshops equipment a number of moulding machines, lathes, boring machines drills and riveting machines-all up-to-date quick speed plant."

A review of all products was taken. This resulted in:

a) Improved and new machinery and implements for the home market;

b) A new range of machinery specifically for export.

A forty page catalogue was issued and a London Office opened at 34, Fenchchurch Street, E.C. 3.

The result of all this was that within sixteen months Corbett-Williams had more than doubled their turnover.

Corbett-Williams were soon exporting to over twenty countries and I quote from The Implement and Machinery Review October 1911:.

> *"Shipments made to:*
> *France , Holland, Belgium, Italy, Denmark, Greece, Turkey, India, the Straits Settlements, Australia, New Zealand, the West Indies, Cape Colony, the Orange River Colony, British East Africa, Natal, Mauritius, Brazil, Peru, Uruguay, and the Argentine Republic."*

Rhuddlan station became a hive of activity as machinery was loaded for both the home and export market. Goods for export would most probably be placed in timber crates and destined for the docks at Liverpool.

Rhuddlan moves to England!

Rhuddlan Station

Rhuddlan Station.

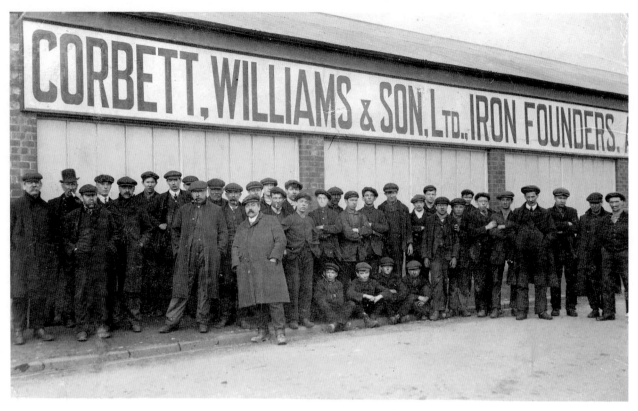

Photo circa 1920.

Credit: Mr & Mrs John Dod (Frank Jones)

Men at 14-15 years of age.

Part 2

Corbett-Williams & Son

1908-1923

Chapter 7

NEW MACHINERY - NEW MARKETS (EXPORTS)

CORBETT-WILLIAMS IMPLEMENTS

October 1911

New Hand Powered Chaff Cutter

Specially designed for export
PICTURE (India) Awarded a silver medal at the 1911 Allahabad Exhibition India
Capable of being packed into a small space for export.

This model a new version of a machine that had been in production for almost fifty years and thousands were sold to the British Army and exported to the Colonies.

At the 1909 Smithfield Show a mutilated steel spanner formed an interesting object lesson as to the strength of the Williams chaff cutter.

This was accompanied by a letter:
> *"I am sending you a spanner which accidentally got in my chaff cutter. I thought by the noise it would all be smashed to pieces but when I had examined it I found only the Knives were knicked a little and after filing them we started again".*

This is also excellent testimony to the quality of the knives which we are informed came from Messers. W. A. Tyzack & Co. Stella Works Sheffield.

March 1912 orders for sixty chaff cutters for South Africa.

New Domestic Grinding Mill

Featured in The Implement and Machinery Review Oct 1911 The New Domestic Grinding Mill. Orders from South Africa, India and South America.
Five models available Mark 0–4.

Corbett-Williams
Domestic Grinding Mills

For Hand Power.

Mark: No. 0.—This Mill can be fixed to a rail or post, or fastened down on a table.

Price **15 /-**

Mark: No. 1.—Suitable for grinding into fine meal for domestic purposes.
Price **£1 12 0**

Mark: No. 2.—Suitable for Kibbling or Coarse Grinding.
Price **£1 12 0**

The above Mills can be fastened down on a Table or Bench.

These Mills have been specially designed to meet the requirements of the Settler in the Colonies and Abroad, where it is necessary to grind grain fine enough for domestic purposes. They are suitable for Farms, Stables, Grocers, Chemists, or Poultry Feeders.

They will grind into fine, coarse, or moderately fine meal, or kibble all kinds of Grain, Coffee, Chicory, Cocoa, Charcoal, Drugs, Rice, Pepper, Spices Linseed, Wheat, Maize, Beans, Peas, Barley, Oats, Mealies, and will grind into the finest powder Chemical, Vegetable, or Mineral substances, broken Oyster Shells, Bones, Dog Biscuits, and Oil Cake.

1912 New Corn and Malt Crusher

Corbett-Williams
NEW PATTERN IMPROVED
Corn and Malt Crusher.

This design is the result of many years' experience in Grinding and Crushing Machinery, and embodies all the modern improvements.

Patent "Phoenix" Harrows for the foreign and colonial markets

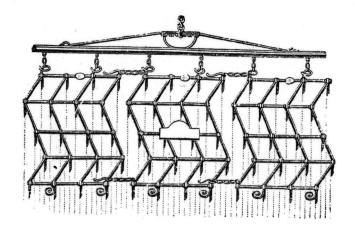

Steel Frame Horse Hoes
The Corbett-Williams New Expanding Horse Hoe of 1912

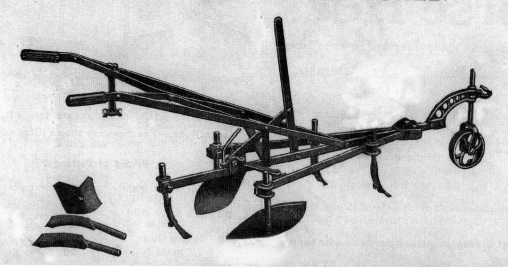

Made of steel and malleable iron thus combining strength and rigidity. The stalks of the tines are round and can be set to any desired angle. The width of the implement may be instantly adjusted by a lever and hand screw.

Weight 100lbs it sells for £2 7s 6d a furrower is 10s 6d extra.

Two hundred were ordered from Ireland with large numbers going to South Africa some agents taking forty-eight at a time.

Mealie (Maize) Planter
The Corbett-Williams Mealie (Maize) Planter

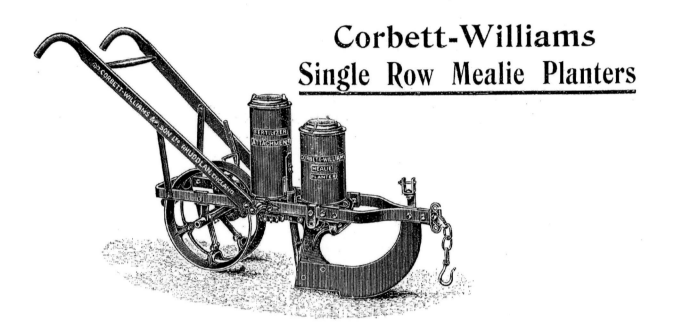

First featured in the Implement and Machinery Review March 1st 1911.

"Designed specially to meet the requirements of Colonial husbandmen, to whom it is capable of affording very considerable assistance. It is we hear being largely shipped to South and East Africa and India."

The machine is provided with a complete set of delivery plates perforated with holes of a suitable size for all grades of maize. There is also two sets of plates for Kaffir and other seeds. It is claimed that it will sow maize at 10, 15, or 20 inches apart.

Turnip and Mangold Drill The "New Era"

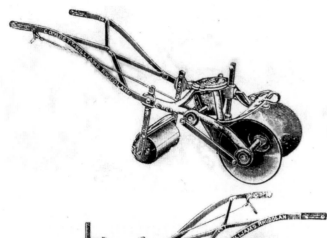

The Corbett-Williams
NEW IMPROVED
Turnip and
Mangold Drill
The "New Era."

STEEL SPINDLES AND FRAMING THROUGHOUT.

Special Note to Colonial and Foreign Buyers.

Where it is the custom to sow on the flat instead of on the ridge, we supply this Drill fitted with two wheels in place of the front concave roller, and with the flat roller for covering the seed fitted at back.

Price £2 - 0 - 0

A New Machine of Unique Design—Practical, Efficient, and Well-made.

Price			
With One Concave Roller £2 0 0	..	(Telegraphic Code—**Era**).
With Wheels in lieu of Front Roller.. £2 0 0	..	(Telegraphic Code—**Erasmic**).
With Two Concave Rollers 2 8 6	..	(Telegraphic Code—**Erat**).
With Wheels in lieu of Rear Roller 2 8 6	..	(Telegraphic Code—**Erratic**).

This Machine is provided with a Handpiece under the Handle, so that the seed can be instantly stopped when turning at the headlands. The driving cranks are fixed on the spindles with cross pins, and cannot work loose, no nuts being used.

The "New Era" had been in production since 1895.

INSTRUCTIONS

FOR THE PROPER REGULATION OF DISCS AND STOPPERS OF

THE CORBETT-WILLIAMS "NEW ERA" DRILLS

For the correct Sowing of the quantities per acre, and for stopping delivery of Seed at the headlands.

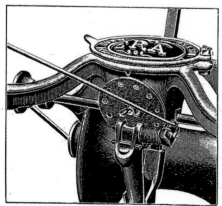

Fig. 1.

In Fig. 1 is shown an enlarged view of the regulating disc and stopper, the latter acting as a cover to the seed pipe.

Fig. 2.

In Fig. 2 the stopper is shown thrown forward, thereby preventing any possible escape of seed.

SEED REGULATOR DIAL: Can be **quickly** and **accurately** adjusted to sow any desired quantity of seed per acre. The letters **T** stamped on the brass dial mean **Turnips**, and the letters **M** mean **Mangolds**. If it is required to sow 5lbs. of Turnip Seed per acre, the mark **T 5** is brought up to the small pointer at the top when the desired quantity per acre will be sown, and so on with other quantities as required.

The Seed can be seen coming through the Disc when working.

The CW Root Cutter and Cleaner Rev 1911 Royal Show at Norwich

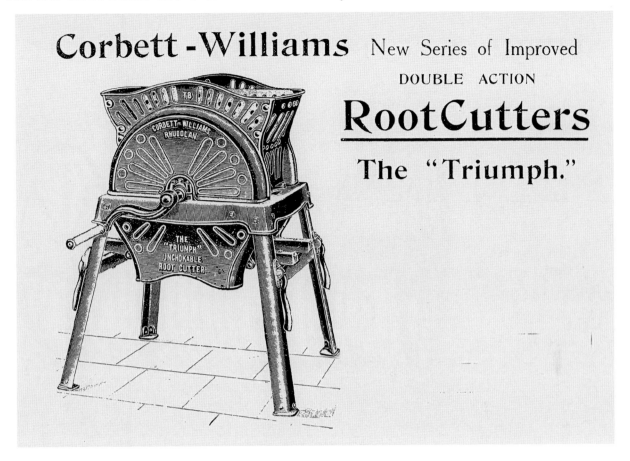

This machine won a silver medal from the Vale of Clwyd Agricultural Society.

Based on their well proven "Triumph Unchokable Root Cutter" it is reported in The Implement and Machinery Review July1, 1912 and exhibited at the Royal Show held at Doncaster.

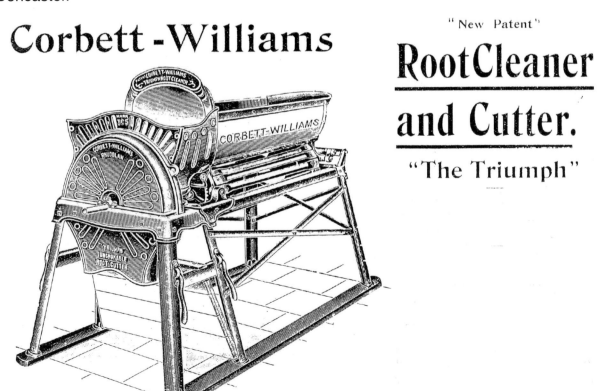

Corbett-Williams Horse Rakes

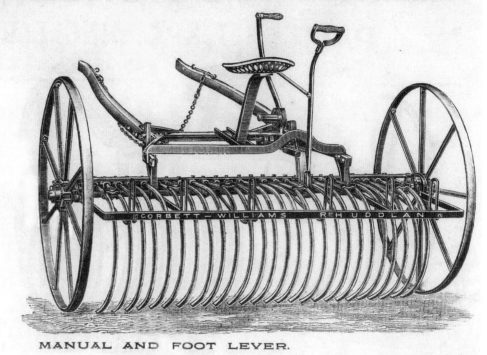

Corbett -Williams "Victory" Grinding Mill

(Patent Francis Corbett) On March 1st 1912 Williams received an order from South Africa for fifteen mills.

In 1920 the "Victory" grinding mills were made in three sizes.

Grist Mill
Credit: Gwynfor

Higham's Gold Ore Samplers 1911 Sole makers Corbett Williams

The machines destined for South Africa and take a continuous sample of the slurry as it laves the stamps.

A GROUP OF HIGHAM'S GOLD ORE SAMPLERS FOR SOUTH AFRICA.

1920 The Corbett Williams "Queen" Mower and Reaper

The Corbett-Williams "Queen" Mower.

A MARVEL OF STRENGTH AND SIMPLICITY.

NO SIDE DRAFT.

THE QUEEN

Roller Bearings.
Spring Assisted Foot Lift.
Light Running.
Close and Even Cutting.

CORBETT-WILLIAMS

Price—As a Mower .. .. £16 0 0   As a Reaper £18 15 0

Horse Gears

These machines had been in continuous production for almost fifty years but with the development of oil engines their days were numbered.

At the Royal Show circa 1912.
left F.A. Goldsmith, ? ,? ,?, David Jones, ,??, Iorwerth Jones, young man.
Credit: Mr & Mrs John Dod (Frank Jones)

Credit: Mr & Mrs John Dod (Frank Jones)

Part 2
Corbett-Williams & Son
1908-1923

Chapter 8
THE FIRST WORLD WAR
1914-1918

At the outbreak of The First World War British exports of agricultural machinery ran at an all time high. Many companies became dynasties and many founding fathers of the industry purchased large country estates. However, war and its aftermath was to be disastrous for the industry and Corbett-Williams fared badly.

Credit: Mr & Mrs John Dod (Frank Jones)

On the eve of The First World War

On August 13th 1913 Francis Corbett was gazetted to a commission as a Second Lieutenant in the 5th Royal Welch Fusiliers, who at the outbreak of war conducted a vigorous recruiting campaign which resulted in many from the district joining the Colours. Almost half of the workers left for military service and women replaced them at the lathes and machines.

By the end of 1914 the government arms factories could not keep up with the demand for munitions. In May 1915 David Lloyd George as Minister of Munitions realized the urgency of the situation and contracts of enormous value were issued to engineering firms.

The premises of Corbett-Williams became engaged in the production of munitions (shell casings) and became a Government Controlled Establishment.

In April 1917 the Ministry of Munitions ordered all controlled establishments to pay an extra 6 shillings bringing wages to 25s a week. Added to this was the rapid rise in cost of raw materials.

The new government under Lloyd George formed the Food Production Department whose object was to bring an extra 3 million acres "under the plough".

With the loss of horses and men to the war what was needed now were the new tractors, ploughs, binders and threshing machines.

The money the farmer had made (rise in price of farm produce of 300%) was likely to be spent on the mentioned machinery.

Companies like Corbett–Williams with their range of horse and long lasting barn machinery were now in the cold.

David Jones

Evans Pen y Bont

William Williams

Harry Griffiths (Joint Foreman)

Iorwerth Jones (Joint Foreman)

Tommy Evans (Undertaker)

Mrs Rees (nee Evans sister) Penybont

Mrs Thomas Cross St

1st Portion Munition Workers — Rhuddlan 1915 –16 —

Credit: Mr & Mrs John Dod (Frank Jones)

Munition Workers — Rhuddlan Iron Works 1915

Capt Francis Corbett and women munition workers.

Credit: Mr & Mrs John Dod (Frank Jones)

Capt Francis Corbett and women munition workers.

Credit: Mr & Mrs John Dod (Frank Jones)

Aftermath of War and Loss of Markets

Corbett-Williams had built a healthy export trade with India, S Africa, S America and Australia, but shortage of shipping, the U-boats and the subsequent rising costs were serious blows. Rhuddlan resumed production of oil engines but the competition was fierce.

The North American companies such as AMANCO Massey-Harris, International Harvester Fairbanks Morse were now shipping mowers reapers and engines into Britain. In market towns all over Wales ironmongers became agents for American engines and machinery.

Credit: Mr & Mrs John Dod (Frank Jones)

Part 2

Corbett-Williams & Son 1908-1923

<u>Chapter 9</u>

Iorwerth Jones Joint Works Foreman and Foreman of "The Oil Engine Shop"

Career

1901–10	John Williams & Son Rhuddlan.
1910–12	Petter & Sons Yeovil.
1912–23	Corbett- Williams & Son Rhuddlan.
1924–1927	Hesford & Co Ormskirk.
1927	Coal Merchant Rhuddlan.

The brothers Iorwerth and David Ellis Jones of Blair Mont Rhyl Road Rhuddlan were gifted young men.

On October 21st 1901 Iorwerth aged fourteen started his apprenticeship at John Williams & Son. He was followed by David his younger brother of four years.

They probably brought into the works a vision of a new age when horse and steam power would be eclipsed by the fledgling oil engine.

The "Jones Brothers" were probably instrumental in persuading Williams to move in this direction.

During 1908 much of their time at the works would have been occupied by the "Oil Engine Project".

The Implement and Machinery Review dated January 1st 1909 made an important announcement: the purchase of the business by Francis Corbett (Corbett- Williams). See next chapter The "Corbett-Williams Oil Engine".

Since 1896 JB Petter & Sons Yeovil Somerset had been making oil engines ranging from 1 ½ to 50hp. They employed hundreds making engines many destined for export.

On July18th 1910 aged twenty-two Iorwerth moved to Yeovil and started work at the mighty Nautilus Works.

The photo shows Iorwerth sitting near an experimental V (vertical) engine indicating that he worked in the development and testing shop.

Having gained valuable experience in 1912 he returns to Rhuddlan. On March 18th of that year he was appointed Joint Works Foreman and Foreman of the oil engine shop.

Photo taken circa 1911 at Petters of Yeovil, Iorwerth seated on the left.
Credit: Mr & Mrs John Dod (Frank Jones)

In 1914 at the outbreak of the War Iorwerth married Margaret Williams at Ebenezer Calvanistic Methodist Chapel Rhuddlan.

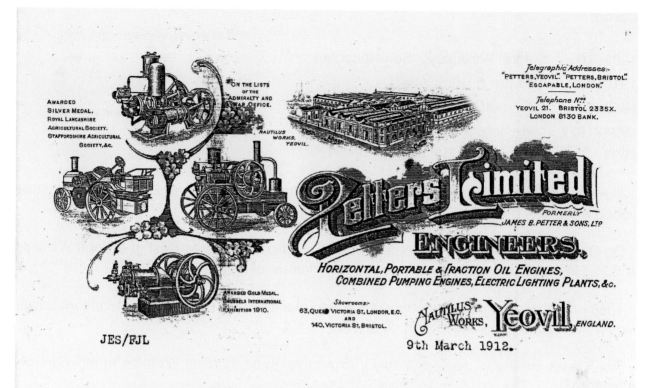

(See The First World War)

Probably a sad time as oil engine development ceased and production given over to munitions.

On cessation of hostilities oil engine production resumed (See Chapter 10 Oil Engines)

Iorwerth Jones Joint Foreman
Blair Mont

David Jones Central Garage

David Jones, Harry Griffiths (Joint Foreman), Iorwerth Jones (Joint Foreman),
Evans Pen y Bont

Credit: Mr & Mrs John Dod (Frank Jones)

Iorwerth at the Royal Show circa 1912.

Credit: Mr & Mrs John Dod (Frank Jones)

Credit: Mr & Mrs John Dod (Frank Jones)

When Corbett-Williams closed in November 1923 Francis Corbett gave Iorwerth a glowing testimonial.

Contractors to His Majesty's Government and the Crown Agents for the Colonies. On the Admiralty and War Office Lists.

TELEGRAMS:
"PHŒNIX, RHUDDLAN."
CODE 5TH EDITION A.B.C.
WESTERN UNION CODE.
TELEPHONE Nº 4 RHUDDLAN.
LONDON OFFICE:
11. BILLITER SQUARE, E.C.3.
TELEGRAMS: CORBETT, GALBANEOS, LONDON
TELEPHONES: AVENUE 295 & 296.

PHŒNIX IRON WORKS
RHUDDLAN,
FLINTSHIRE.

Corbett, Williams & Son,
AGRICULTURAL IMPLEMENT
LIMITED
MANUFACTURERS
AND ENGINEERS.

WINNERS OF over 300 FIRST PRIZES, GOLD and SILVER MEDALS, for SUPERIOR IMPLEMENTS and MACHINERY.
AWARDED FIRST PRIZE AND SPECIAL SILVER MEDAL AT THE ALLAHABAD (INDIA) EXHIBITION.
Manufacturers of the "CORBETT WILLIAMS" OIL ENGINE.   1st. November 1923.

I have much pleasure in stating that Mr. Iorwerth Jones served his apprenticeship at the above works, commencing on October 21st. 1901. and remaining with us until July 6th. 1910. He served his apprenticeship in our General Engineering Shops, and had very good experience in Repairs to Steam & Traction Engines, and in the manufacture of Internal Combustion Engines.

He gave the greatest possible satisfaction, in every way, and became a highly skilled Fitter and Erector. He was always most civil and obliging, and his work was at all times reliable, and good. He left us in 1910 for a short period to gain further experience, and re-started here on March 18th. 1912 and is still with us at the present date. On his return, he was appointed Foreman of the Engine Shop, and was responsible for the erection and testing of Oil Engines for the War Office, Marine Motors for the Admiralty, also Petrol Engines, Petrol-Paraffin Engines, and Horizontal Oil Engines which we have shipped to all parts of the world.

He is a most reliable, experienced, efficient, straightforward, willing hardworking Foreman, a good time-keeper, liked by the men, and a good disciplinarian.

I have much pleasure in giving him this Testimonial, and may say it has always been a pleasure to work with him.

Managing Director.

Credit: Mr & Mrs John Dod (Frank Jones)

Enclo⁵_____

ARICULTURAL MACHINERY AND FARM TRACTOR SPECIALISTS.

Telephone Nº 64   TelegraphicAddress: HESFORD, ORMSKIRK.

CHAS. M. HESFORD & CO.
(C.M. HESFORD SOLE PROPRIETOR.)

Engineers & Ironfounders,

Ormskirk.

In reply please
quote this Reference

January 8th. 1927.

The Bearer, Mr. Iorwerth Jones, was

employed by us as working foreman engine fitter, and

tester, from August 1924, to January 1927 , during this

period he gave us every satisfaction, and regret that

owing to domestic affairs, we are losing his services.

We shall be pleased to answer any enquiries

And would recommend him to any firm with confide-

-nce.

CHAS. M. HESFORD & CO.

Robert. C. Rose.

.Works Manager.

Credit: Mr & Mrs John Dod (Frank Jones)

From August 1924 until January 1927 Iorwerth worked as the "oil engine man" at Chas. M. Hesford & Co Engineers and Ironfounders of Ormskirk Lancashire.

It appears that his wife and family remained at Blair Mont Rhuddlan.

On his return to Rhuddlan he started up in business as a Coal Merchant, his brother David becoming proprietor of the Central Garage.

Iorwerth passed away aged 57 years on June 2nd 1945.

David Jones passed away on February 4th 1959 aged 83. His two sons were Len and Wilf (Central Garage).

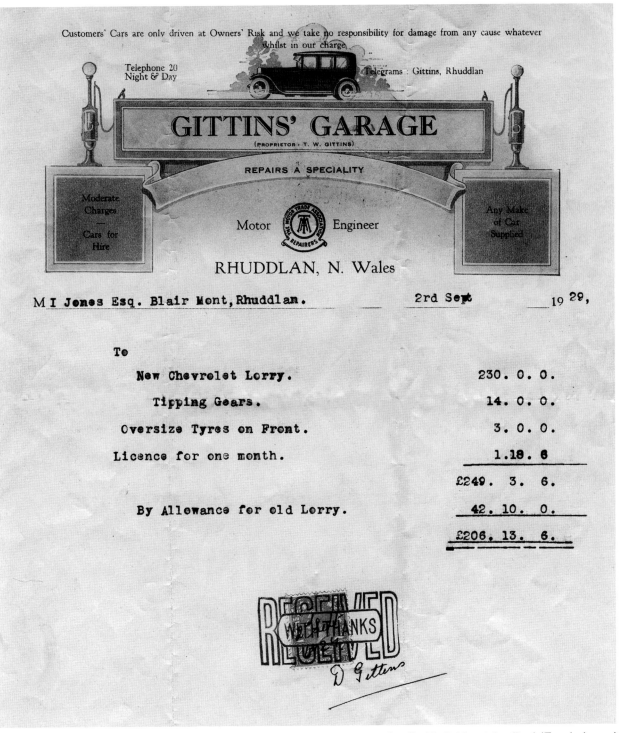

Credit: Mr & Mrs John Dod (Frank Jones)

Part 2

Corbett-Williams & Son 1908-1923

Chapter 10

OIL ENGINES

To quote from *A History of Wales* by John Davies:

"The only sources of power were the strength of men of men and beasts, the power of wind and water none of these was tireless or wholly reliable".

As previously mentioned feed had to be prepared fresh every day. The problem for the stockman lay in the lack of a cheap and above all a simple and reliable source of power.

In winter when the demand for feed was high, snow and frost could severely add to the problem.

My mother as a young girl on the farm of Maestyddyn Isa Clawddnewydd used to recall the waterwheel frozen solid and the stockman would come into the house and ask for "all hands on deck in the barn".

Imagine also under such conditions the problem of hitching the horse or horses to the horse gear and working it.

In Wales only a few large land owners might have had steam power.

The answer started to emerge during the 1880s and lay in the development of the Oil Engine, and by1908 there were some 13,000 in use on farms. These engines, of course, tended to be on the larger farms and made by companies that specialised in engines names like Rushton Hornsby, Blackstone and Crossley. Some of these early engines once started were more suitable for continuous running under relatively clean conditions, i.e. factories, pumping and electricity generation. Note I have used the word once started! With an Oil Engine running at full power for half an hour all daily feed could be prepared on most farms.

Companies like John Williams with years of experience making farm machinery took a basic realistic approach to Oil Engines. They set out to design and develop engines almost specifically for powering barn machinery.

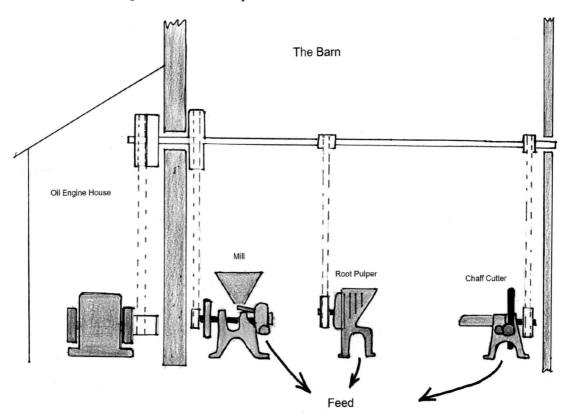

Engines

This was the first time that most farmers had come into contact with an internal combustion engine. Manufacturers produced booklets under titles such as Reducing Costs extolling the advantage of their Oil Engines over horses or hand labour.

"Compared with a horse gear The Engine does the work 50% faster and the horses are available during the busy season for other work. You also save the time in getting out the horses harnessing them to the gear and looking after them as they go round.

Manufacturers stated that their engine was:
Easily Started
Easy to Run
Easily Maintained (by un-skilled labour)

The stock farmer is not renowned for his interest in machinery but to be fair the majority managed pretty well.

A plethora of Oil Engine manufacturers appear on the market place many producing Engines from 11/2 H.P .to say 10 H.P. at very affordable prices.

By about 1908 William Bridge Williams at Rhiuddlan would be pondering over the risks and rewards of Williams producing their own Oil Engines.

Rhuddlan had hitherto gained a reputation for the excellence of their machinery so an engine bearing the "Williams" name should sell.

Williams had dealerships throughout Britain Ireland and the Colonies that would handle the new product.

What better at the Royal Show than have on your stand barn machinery powered by your own engine.

Having an Oil Engine at this time meant that you were sending a message that you are a progressive company embracing the latest technology.

The first Williams engine of 1908-1909

This engine bears a strong likeness to an engine made by Charles Price & Sons Broadheath Manchester (see A–Z of British Stationary Engines Volume Two L–Z Patrick Knight)

There could have been a reciprocal agreement where Williams made the castings and Price made the engines, of this we shall never know!!

THE CORBETT-WILLIAMS OIL ENGINE.

January 1st 1909 Implement and Machinery Review

"No little interest appears to have been bestowed upon their new Corbett-Williams oil engine, regarding which a telegram was received just before our arrival recording the behaviour of the first they have actually placed. This motor runs on paraffin after it has been started on petrol, a third of a pint sufficing for that purpose".

Ignition appears to be by means of a low tension magneto and igniter, a system common to many engines of this period.

It appears that none of these engines have survived into preservation.

Corbett-Williams Horizontal Oil Engines (Lamp-Start Hot-Tube Ignition)

First introduced in 1909 with the words:
"It's first cost is reasonable, the maintenance cost is nominal, the working cost is fractional. Such recommendations were advanced as an inducement for visitors (to the Smithfield Show) to examine the 6 ½ hp engine shown by Corbett-Williams".

Corbett-Williams Oil Engines

THE MOST RELIABLE, ECONOMICAL, AND SIMPLE ENGINE MADE.

Basically, a blow lamp is lit and the tube heated for about five minutes. On being sprayed on to the tube the fuel ignites. The lamp can then be extinguished and the engine should keep on running. To keep the tube hot, this type of engine prefers to run under fairly heavy loads at about 250–350 rpm. It tended to find favour on the larger farms where it would run for longer periods.

The smallest model of 4 hp weighed well over half a ton. It consumed about ¾ of a pint of fuel per bhp/hour.

The 1913 Royal Show was held at Bristol and on the first day, forty oil engines were ordered and were to be sent to their newly opened branch office in Melbourne Australia.

PARTICULARS

OF THE

"Corbett-Williams" Horizontal Oil Engines.

ORDINARY TYPE FOR REFINED OIL (PETROLEUM, PARAFFIN, or KEROSENE.)

MAXIMUM. B.H.P.	Net Weight Appr'x	Gross Weight Packed for Shipment Appr'x	Approximate Cubic Measurements Without Tank and Pipes	Complete	Space occupied by Engine only		Net Weight of Standard Portable Engines	Approximate Cubic Measurements Standard Portable Engines	Standard Size of Pulley	Size of Fly-wheels	Speed Revolutions per Minute	MAXIMUM. B.H.P.	PRICES. Without Tank and Pipes	Complete with Tank and Pipes	Standard Portable
	Cwts	Cwts	Cub. ft	Cub. ft.	Ft. Ins.	Ft. Ins.	Cwts	Cub. ft.	Ins.	Ft. Ins			£	£	£
4	10	13	34	55	4 9	x 2 8	23	190	10 x 6	2 6	360	4			
4½	10½	13½	38	59	4 10	x 2 9	23½	200	10 x 6	2 6	360	4½			
5½	14	17	40	60	5 1	x 3 4	29	210	12 x 6	2 8	350	5½			
6½	15	18	45	65	5 3	x 3 6	30	220	12 x 6	2 10	340	6½			
8½	22	28	60	100	6 2	x 4 5	35	220	16 x 9	3 6	320	8½			
10	22½	29	65	105	6 4	x 4 5	37	225	16 x 9	3 6	300	10			
11½	23	30	70	110	6 6	x 4 6	39	230	16 x 9	3 8	280	11½			
13½	28	35	75	115	6 8	x 4 7	45	240	18 x 10	3 8	275	13½			
15½	36	44	100	176	7 3	x 4 8	52	320	20 x 12	4 2	260	15½			
17	40	49	110	186	7 4	x 4 10	56	350	20 x 12	4 2	250	17			
19	44	53	132	208	8 1	x 5 7	60	380	20 x 12	4 4	250	19			
21	57	67	134	210	8 2	x 5 8	70	460	22 x 12	4 6	240	21			
23½	60	71	183	263	8 5	x 6 0	72	480	22 x 12	4 6	240	23½			
26½	63	75	210	290	8 9	x 6 1	75	500	24 x 13	4 9	240	26½			

All Engines are fitted with Two Fly-wheels.

ALL MEASUREMENTS, DIMENSIONS, AND WEIGHTS GIVEN ARE APPROXIMATE.

THE CORBETT-WILLIAMS OIL ENGINE.

The "Corbett-Williams" Portable Oil Engine

THE "CORBETT-WILLIAMS" PORTABLE OIL ENGINE.

Particularly useful when work has to be done in different locations.

At the Royal Show of 1914 held at Shrewsbury Corbett-Williams reported that they had sold the 7 ½hp, 10 ½hp and16hp engines shown on their stand and that they had a contract with one firm abroad to supply 100 engines per year.

The First World War

The Royal Agricultural Show held at Nottingham in1915 was to be the last until the show at Cardiff in 1919.

At Cardiff The Implement and Machinery Review reported: "Whether oil-boring in this country fulfils anticipations or not, there will still be a big run on oil engines."

Confirmation of this view we found on the stand of Messers. Corbett-Williams & Son. Ltd., Rhuddlan, Flintshire. There was a particularly good display of stationary and portable oil engines ranging in power from 4 ½ to 22 b.h.p.

The "Corbett-Williams" Portable Pumping Plant

THE "CORBETT-WILLIAMS" PORTABLE PUMPING PLANT.

These pumping plants were built for export to Egypt India and Australia.

Made in six sizes from 4 ½ to18hp they will pump from 4,200 to 54,000 gallons per minute with a total lift of 25 feet. The pump is driven direct from the engine flywheel, fast and loose pulleys being fitted on the pump spindle. Several of these sets were sold to the Hunter Valley irrigation scheme in New South Wales.

Two types of sets

Engine cooling water can be either (a) taken from the pumping plant or (b) from the engines independent system, so the engine can be used for purposes other than pumping.

The "Corbett-Williams" Petrol Engine (The Lister look-alike)

"Corbett-Williams" Vertical Petrol Engines

3 H.P. and 5 H.P.

Starts instantly. No lamp required. Complete with High Tension Magneto Ignition, Water Tank and Standard Connections, Silencer and Standard Pulley.

Stationary Type.

Brake H.P.	Speed R.P.M.	Flywheel diam.	Driving Pulley.	PRICE.
3	450	22″	8″ x 5″	£
5	450	24″	10″ x 6″	£

"Corbett-Williams" Portable Petrol Engines.

Mounted on Strong 4-wheeled Wooden Trolley with Cooling Tank.

Portable Type.

Brake H.P.	Speed. R.P.M.	Flywheel diam.	Driving Pulley.	Price with Ordinary Standard Cooling Tank.	Price with Square Cooling Tank and Circulating Pump.
3	450	22″	8″ x 5″	£	£
5	450	24″	10″ x 6″	£	£

N.B.—The Square Cooling Tank and Circulating Pump are lighter for transport than the Ordinary Cooling Tank, and, moreover, where there is a limited water supply this arrangement is a distinct advantage.

The engine machine-shop at Rhuddlan.
Note a man sitting on what appears to be an engine cylinder.

Credit: Mrs F Roberts, Trefnant

Engine machine shop.

Introduced at the Darlington Royal Show 1920

These engines bear a strong semblance to the early petrol engines built by R. A. Lister & Co Dursley, Glos.

A vertical enclosed crankshaft engine available in two power sizes 3 and 5hp, it will also run on paraffin after starting up on petrol.

The design and physical size of both models are the same. The only external difference is that the 5hp engine has larger/heavier flywheels and driving pulley.

At the 1922 Royal Show held at Cambridge

The "Corbett Light" was introduced. A tank-cooled 3hp paraffin Williams engine turning at 450 r.p.m coupled to an English made dynamo by belt and mounted on a sturdy cast-iron bed plate.

Corbett-Williams Engines in Preservation

At the time of writing there are about fourteen known engines. Seven reside in Australia with four in Wales and two in England.

Oil Engines-about fourteen, of which seven are known to be in Australia all in private collections. Working on an estimate that only two out of every 100 engines remain, then the total number of engines produced at Rhuddlan would hardly reach 1000. This would have been over a period of fourteen years, just above one engine made per week!!

3 H.p. Owned by Karl Schroder of Eglwys Cross.
Photo by author at Dyffryn Iâl Vintage Machinery Rally held at Ruthin 2014.

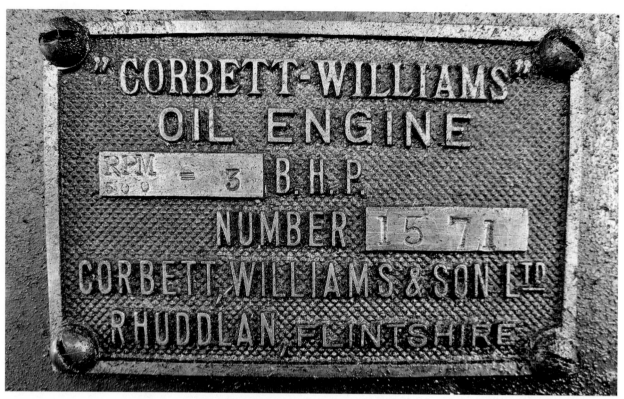

Nameplate of previous engine.

Credit: Gwynfor

6 H.p. Owned by Eifion Jones.

Credit: Gwynfor

6 H.p. Owned by Reg Davies Rhuddlan.

Credit: Gwynfor

Horizontal Engines
Australia
No 782 Hp 7 ½.
No 813 Hp 3 ½ .
No 861 4 ½ Hp.
No 864.
No 1019 Hp 4 .
No 818 Hp 4 ½
Wales 3
No 1000 Hp 4 ½.
No 1002 Hp 4 ½ .
Another?

Vertical Engines
Wales
No   5 Hp
England
No   5Hp
No 1571 3Hp.

Not far from home this 4 ½ H.p. lamp start shown by John Paterson of Dyserth.

Credit: Gwynfor

Engine enthusiast Hans Jensen of Benalla, Victoria, Australia bought the remains of a 4Hp Corbett-Williams No1019. Many parts were missing not least the crankshaft and flywheels. After several e-mails to John Paterson of Dyserth Nr Rhuddlan the dimensions of the missing parts was worked out.

Many hours of patience and skill has resulted in an engine that "runs as well as it looks".

Far from home this 7 ½ H.p. portable lies in Australia.

Credit: Hans Jensen

Nameplate yet another Corbett-Williams found in Australia.

Credit: Hans Jensen

"That is all I have of this 4 ½ H.p. CW engine" Hans Jensen.

Credit: Hans Jensen

*Hans Jensen,
Benalla, Victoria, Australia.
Ex state of Victoria road builder backyard
fiddler. One of Australia's most respected
engine restorers.*

Credit: Gwynfor 2012

*"Crank and flywheel of a Keighley made
Pitt may do the job"*

Credit: Hans Jensen

Parts manufactured and machined.

Credit: Hans Jensen

The finished job, probably the finest restored Corbett-Williams in the world.

Credit: Hans Jensen

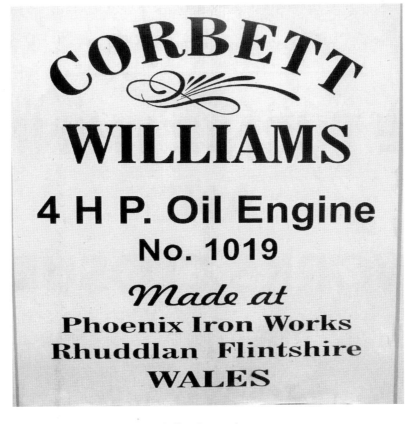

A final touch.

Credit: Hans Jensen

Part 2

Corbett-Williams & Son 1908-1923

Chapter 11

THE TURNIP THINNER AND WORKS CLOSURE

The Corbett-Williams Root Thinner

Side view of machine.

Any reader having experienced thinning and singling rows of turnips will know what soul destroying work it was.

This machine in single row version was exhibited at the 1920 Darlington Royal Show under the name of Mr Harold William Davey, Maes Mynan Hall, Afonwen, Flintshire.

Field Trials of Corbett-Williams & Son Turnip Thinner

Photo dated 23rd of February 1923. Maes Mynan Hall Afonwen Nr Denbigh.

Held at Maes Mynan on February 23rd 1921. A two row version had been developed and the trial was witnessed by a large number of influential agriculturalists and implement agents who came from Flintshire, Denbighshire and Cheshire.

The efficiency of the gapping and the labour-saving qualities were noted.

Rear view of machine.

At the 1921 Derby Royal Show the machine was also exhibited in two row version.

The Judge Harry W. Buddicom, Penbedw, Nannerch N. Wales commented that "the machine was much improved but that feed back from farmers would be useful."

It is probable that the "root thinner" was never put into full production.

At the time of writing none are in preservation.

The 1921–2 depression and a collapse of currencies aided by coal and railway strikes meant that Corbett-Williams had no reserves to draw on.

Front view of machine.

At the 1922 Royal Show held at Cambridge the "Corbett Light" was introduced. A tank-cooled 3hp paraffin Corbett-Williams engine turning at 450 r.p.m coupled to an English made dynamo by belt and mounted on a sturdy cast-iron bed plate.

Corbett-Williams did not exhibit at the Royal Show at Newcastle 1923 and closed on November 31st 1923.

The last game.

Credit: Mr & Mrs John Dod (Frank Jones)

"We are closing lads" November 1923.

Credit: Mr & Mrs John Dod (Frank Jones)

Part 3
The Rhuddlan Foundry
1923-2000

Chapter 12

FINAL DAYS

After the closure of Corbett-Williams in November 1923 it is very likely that due to depressed markets the foundry remained dormant for months if not years. Some of the patterns and designs were acquired by their old rivals Powell Brothers at Wrexham.

Mc Lachlan & Co of Armley, Leeds purchased the patterns of the "Oil Engine".

Many buildings on the leased site would have been turned into other uses.

Iball Brothers, a Buckley family re-opened the foundry albeit a shadow of its former glory. It employed about six workers supplying general castings.

In October 1950 a new company was formed named Sterland and Partners, the three directors being Tom Reece, Ben Simpson and J. Sterland.

Circa 1952, J. Sterland was replaced by Jim W.D. Taylor and along with Reece and Simpson formed a new company named "The Rhuddlan Foundry Ltd".

Jim Taylor was also the Managing Director of William Ayrton of Longsight, Manchester who manufactured textile winding machinery for sale worldwide. Castings for these machines were now produced at Rhuddlan.

Despite a fire in the early 1950s, investment was made in a new office, a larger cupola furnace and a process known as shell moulding. This process produced castings of good dimensional accuracy that required little or no further machining. Thus, Rhuddlan became the main supplier of water pump impellers and housings to the Quinton Hazell car component factory at Conway Road, Mochdre near Colwyn Bay. During the 1960s the foundry flourished employing about twenty.

Pride in the melting shop

by CLIFFORD GORDON WILLIAMS

It is natural enough that the countless thousands who pass over Rhuddlan bridge have eyes only for the Castle. Here is history, proud history. The other side of the river, however, a little downstream, is Rhuddlan Foundry which still functions in the same buildings in which it started its activities in the early 1850s. Here pride is in the melting shop.

"We have been here a very long time, says Mr B. Simpson. He and his co-director, Mr Reese, are proud of the contribution they make to industry throughout the country.

The crimson cloud

Employing some 15 to 20 local people the foundry is continuously engaged in the manufacture of castings for the electrical, textile and machine tool industries. Countless numbers of impellors are cast here for the motoring industry and industrial machinery. These find their way all over the world.

Thus a fine contribution is made by this small foundry to the export market.

The scene at the foundry is quite spectacular when casting takes place and we experienced a sense of wonder, heat and uproar notwithstanding. Across from the majestic castle and the peaceful oasis of the church the crimson cloud of smoke from the foundry furnace stirs one to praise the Welsh industrial worker.

IN THE small town of Rhuddlan, dominated by its old Parliament House, church and castle, the casual visitor would be very surprised to find any sign of industry.

Yet Rhuddlan can boast of an old established foundry which has been functioning in the same buildings for over a century. The foundry which makes track components for railways all over the world plays an important part in keeping young people in the town by providing them with opportunities for industrial training.

The co-directors of the foundry, Mr Reese and Mr Simpson, are proud of the contribution made by this small foundry to rail safety and the export market.

In the picture, right, Mr Reese taps the furnace and draws off the mollten metal. Below, Mr Eley Edwards dresses a casting.

Clifford Gordon Williams

The 31st of March the year 2000 saw the foundry close its doors for the last time, thus bringing to an end over a hundred and fifty years of melting and moulding.

In the same year William Ayrton in Manchester ceased trading.

The foundry was demolished in the spring of 2012, however, many of the buildings remain such as the showroom in the form of Morris's Mowers.

Tom Reece at the furnace.

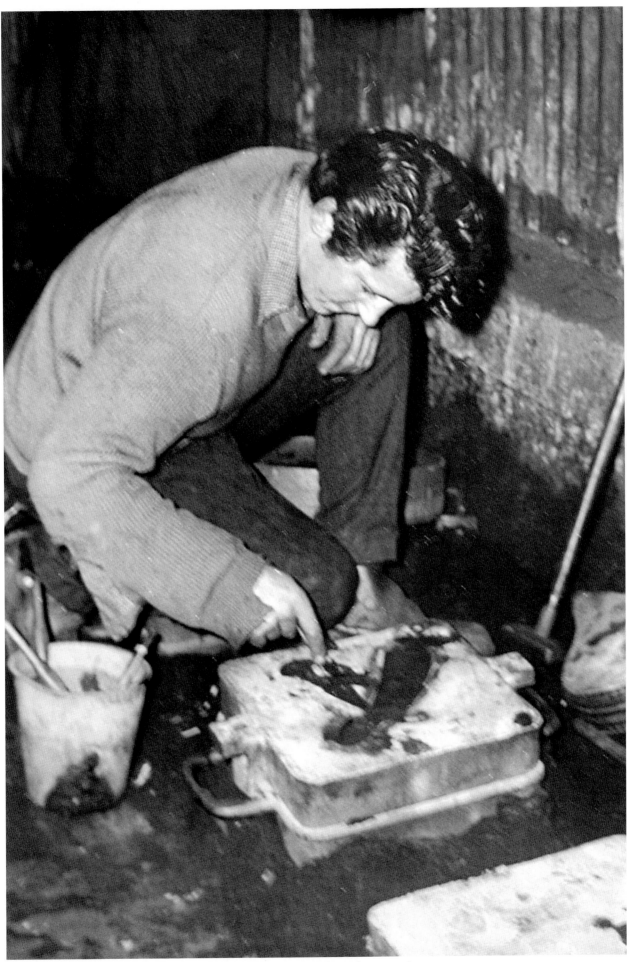

Idris Morris dressing the casting.

Credit: Ted Reece

Tom Reece and Idris Morris pouring into the moulds.

Credit: Ted Reece

The last day Gerard Simpson

Credit: Gerard Simpson

Credit: Gerard Simpson

The last day.

Credit: Gerard Simpson

Closed. Photo June 2000

Credit: Gwynfor

"Going", photo by Ted Reece 9th of May 2011.

Credit: Barrie Mee

"Gone, final demolition" photo by Ted Reece 9th May 2011.

Credit: Ted Reece

Catalogue of Implements and Machinery Circa 1920

Corbett-Williams

MANUFACTURERS OF

High-class Agricultural Implements

and Machinery.

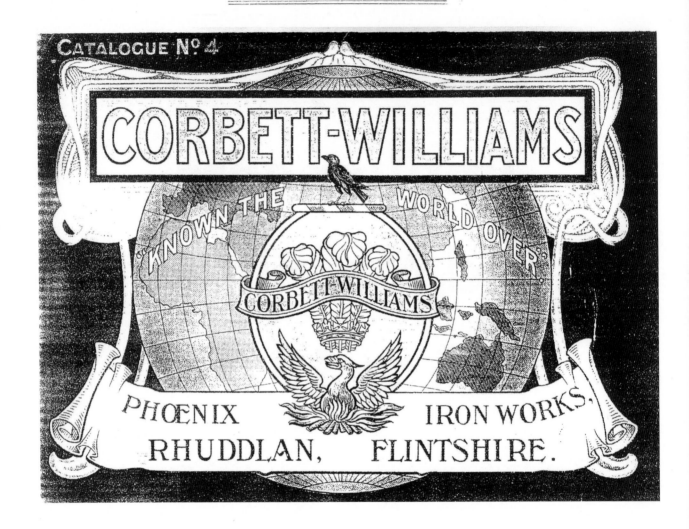

Corbett-Williams

MANUFACTURERS OF

High-class Agricultural Implements

and Machinery.

TERMS.

New Accounts : Approved References or Remittance with order.

DELIVERY.

All orders for Machines amounting in nett value to £2 and upwards are delivered carriage paid to most of the railway stations and ports in Great Britain and Ireland.

When the order is less than £2 in value the carriage in all cases is paid by the purchaser.

All goods should be carefully examined on delivery before signing.

Any loss arising from goods being signed for as complete and in good order and condition, when they are incomplete or damaged, must be borne by the purchaser.

In case any machines or parts are broken **in Transit,** they, or the broken parts, should be returned to our Works at once, labelled with sender's address, and marked—

"CARRIAGE FREE. Broken in Transit."

In every case returning it by the same Railway Company as that by which it was sent.

CORBETT, WILLIAMS, AND SON, LTD., beg to give notice that whilst using every effort to execute orders promptly, sufficient stock being generally held to ensure quick despatch, they must be held free from all liability in the event of any delay in the execution of any order or transit.

All prices, etc., given in this Catalogue are subject to alteration without notice.

Corbett-Williams

Chaff Cutters.

Awarded First Prize and Special Silver Medal at the Allahabad (India) Exhibition held 1910-1911.

NEW DESIGN HAND-POWER

Chaff Cutters.

These Machines are Mounted on Rolled Steel Legs, and have an extra Worm for Cutting two lengths of Chaff.

PRICES :

F 1.	Fixed Mouthpiece, 7½in. wide ..	**£2 2 0**
F 2.	Rising Mouthpiece, 7½in. wide, with Spring	**£2 7 6**
F 3.	Rising Mouthpiece, 7½in. wide, with Weight	**£2 10 0**
F 4.	Rising Mouthpiece, 8¼in. wide, with Spring	**£2 10 0**
F 5.	Rising Mouthpiece, 8¼in. wide, with Weight	**£2 12 6**

Suitable for Small Holdings, and specially recommended for Export.

Corbett-Williams Chaff Cutters.

First Prize Improved New Pattern, Double Cut

Worm Wheel running
in an Oil Bath.

The most efficient and
best made.

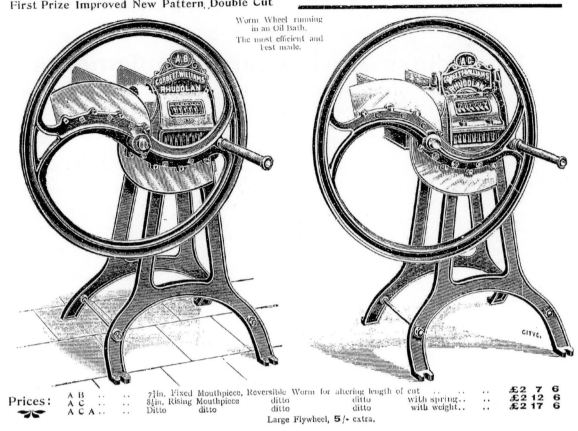

Prices:
A B	..	..	7¾in. Fixed Mouthpiece, Reversible Worm for altering length of cut £2 7 6
A C	..	..	8½in. Rising Mouthpiece ditto ditto with spring.. .. £2 12 6
A C A	..	Ditto ditto ditto ditto with weight.. .. £2 17 6

Large Flywheel, 5/- extra.

Corbett-Williams   NEW DESIGN HAND POWER   Chaff Cutters.

Mounted on Rolled Steel Legs, well braced, Improved Simple Gearing, actuated by Lever Gearing, with neat
oover. Cuts two lengths, ⅜in. and ¾in.

A D, with 8½in. Rising Mouthpiece, **£3 12 6**   A D A, with 8½in. Rising Mouthpiece, **£4 2 6**

Large Flywheels, **5/-** extra.

Corbett-Williams

NEW DESIGN

HAND POWER

Chaff

Cutter,

WITH NEW SAFETY
LEVER.

Mounted on Rolled Steel Legs, well braced, Fly-wheel inside Frame, Brass Bearings, encased Simple Gearing altered by Side Lever. The machine is instantly thrown out of gear by Safety Lever. 9in. Rising Mouthpiece.

A E Price **£5 5 0**   Auxiliary Handle, **10/-** extra,   Hinged Cover, **30/-** extra.

Corbett-Williams

ROYAL PRIZE
SAFETY LEVER
WP
Chaff
Cutters.

FOR
HORSE, STEAM,
GAS, OIL, WATER,
OR OTHER
MOTIVE POWER.

THESE Machines are fitted with Improved Safety Lever, Forward, Stop and Reverse Motions, also with Centre Bearings, making them more rigid for heavy work.

Sent cut to cut ⅜in. and ⅝in. unless otherwise ordered.

– –

The strongest, simplest, and most efficient Cutters made.

Mark.	Price of Machine.	Hinged Covers to Flywheels extra.	Size of Mouth-piece.	Mouth-piece Rises to	Number of Knives.	Speed.		Capacity. ⅜in. chaff per hour.		Engine Brake Horse-Power Required.
						Horse-power.	Engine-power.	Horse-power.	Engine-power.	
WP1	£6 0 0	£1 10 0	9in.	4in.	2	120	150	7½ cwt.	10 cwt.	1½
WP1A	6 7 6	1 10 0	9in.	4in.	2	120	150	7½ cwt.	10 cwt.	1½
WP2	7 7 0	1 10 0	9¾in.	4½in.	2	120	150	9 cwt.	12 cwt.	2
WP3	9 15 0	1 10 0	10¼in.	4½in.	2	120	200	10½ cwt.	15 cwt.	3
WP3A	11 12 6	1 10 0	10¼in.	4½in.	3	120	200	15 cwt.	20 cwt.	3 to 4

Auxiliary Handles, **10/-** each.   Pulleys extra, see page 45.   The Flywheel of the W P 1 A Chaff-cutter is 3in.
Feeding Webs are not fitted to above machines.   larger than that of the W P 1 Machine.

Corbett-Williams Chaff Cutter

New Improved Pattern.

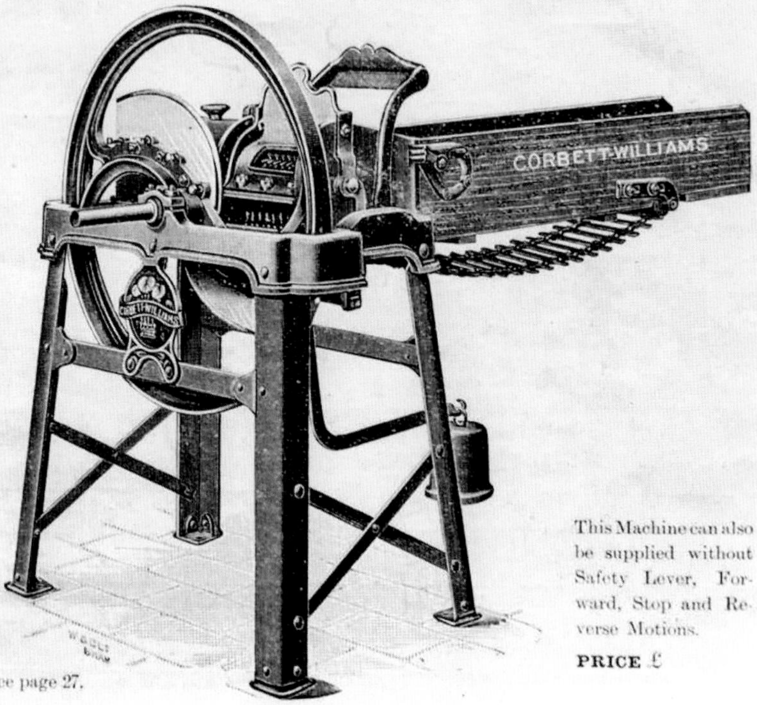

This Machine has been introduced in order to meet the demand for a small power Chaff Cutter, which can be supplied with a travelling Feed Web.

It is fitted with Improved Safety Lever, Forward, Stop, and Reverse Motions, and is arranged to cut two lengths of chaff.

The Gearing is enclosed in a Sheet-steel Cover.

9 inch Mouthpiece, rising from 2¼ inches to 4 inches.

Mark F8. PRICE ... ... £

Extra for Feeding Web ...

Extra for Fly-wheel Cover

Auxiliary Handle, extra. Pulleys extra, see page 27.

This Machine can also be supplied without Safety Lever, Forward, Stop and Reverse Motions.

PRICE £

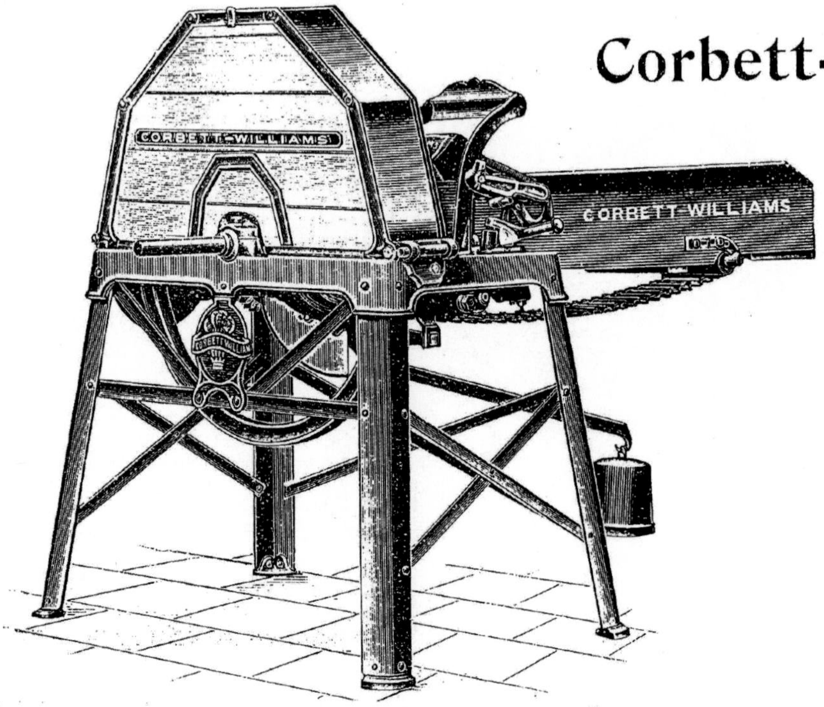

Corbett-Williams

First Prize Gold Medal

NEW PATTERN

Chaff Cutters.

WITH SAFE-GUARD ROLLERS.

No stopping of Machine to alter lengths of cuts, which are changed by hand lever at side.

Machine Marked	Price.	Feeding Web Extra.	Flywheel Covers Extra.	Width of Mouth.	Length of Cuts, Ordinary.	Speed.		Capacity. ⅜in. chaff per hour.		Engine Brake Horse-power Required.
						Horse-power.	Engine-power.	Horse-power.	Engine-power.	
A F	£7 0 0	20/-	30/-	9in.	3/16, 3/8, and 3/4in.	120	150	7½ cwt.	10 cwt.	1½
A H	8 8 0	20/-	30/	9¾in.	3/16, 3/8, and 3/4in.	120	150	9 cwt.	12 cwt.	2
A I	10 15 0	20/-	30/	10½in.	3/16, 3/8, and 3/4in.	120	200	10½ cwt.	15 cwt.	3

Pulleys extra, see page 45. These Machines are fitted with Safeguard Rollers, which rise and fall with top mouth-piece, and are included in above prices.

Corbett-Williams
ROYAL PRIZE
SAFETY LEVER
WP
Chaff
Cutters

For Horse, Steam, Gas, Oil, Water, or other Motive Power.

All these Machines are fitted with Improved Safety Lever, Forward, Stop and Reverse Motions, and with Centre Bearings, making them more rigid for heavy work. Sent out to cut ⅜in. and ¾in. unless ordered otherwise. Pulleys extra, see page 45.

MARK.	PRICE OF MACHINE.	Feeding Web Extra.	Hinged Covers to Flywheels Extra.	Size of Mouthpiece.	Mouthpiece Rises to	Number of Knives.	SPEED.		CAPACITY. ¾in. chaff per hour.		Engine Brake Horsepower Required.
							Horsepower.	Enginepower.	Horsepower.	Enginepower.	
W P 4	£11 10 0	20/-	£1 15 0	11½in.	5¼in.	2	120	200	12 cwt.	20 cwt.	3½
W P 4A	12 15 0	20/-	1 15 0	11½in.	5¼in.	3	120	200	18 cwt.	30 cwt.	4
W P 5	15 0 0	26/-	2 0 0	13in.	5¼in.	2	120	200	20 cwt.	32 cwt.	5
W P 6	16 0 0	26/-	2 0 0	13in.	5¼in.	3	120	200	24 cwt.	36 cwt.	5
W P 7	20 0 0	26/-	2 5 0	15in.	6in.	3	120	220	—	45 cwt.	6 to 8

SLICER KNIVES.

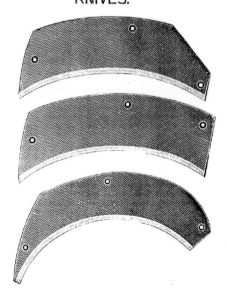

FINGER PIECE KNIFE.

PULPER KNIVES.

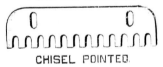

CHISEL POINTED.

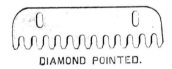

DIAMOND POINTED.

GRATER DISC.

SECTIONS OF SIZES CUT BY GRATERS

EXTRA COARSE FOR SHEEP 1½ INCH

COARSE. 1 INCH

MEDIUM. ¾ INCH

FINE. ⅝ INCH

Corbett-Williams

IMPROVED PATTERN

Single Action 24in. Disc.

Root Cutter

Fitted with Steel Legs.

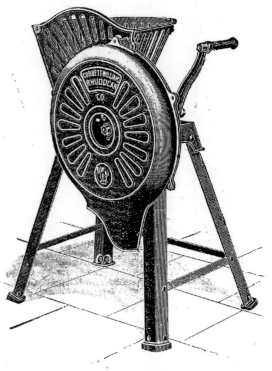

CO	PRICE.
Pulper	
Grater	
Slicer	**£2 15 0**

This Machine can be supplied without Cover
for **£2 10 0**

SIMPLE, DURABLE, STRONG !

WONDERFUL VALUE !

Corbett-Williams

NEW SERIES OF IMPROVED

RootCutters

The "Triumph"

These New Pulpers and Root Cutters are
very strong and substantially made, fitted
with steel legs, unchokable disc, and improved Rotary Feed, and swivel renewable
bearings.

Positively Unchokable.

Pulper, with four chisel point knives	T 1 18in. disc.	T 2 22in. disc.	T 3 24in. disc.
Grater Slicer Finger-piece Cutter	£2 5 0	£2 15 0	£3 0 0

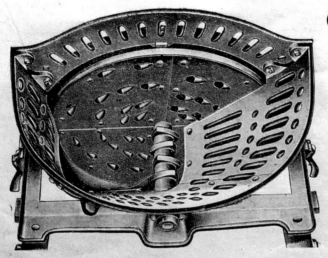

Corbett-Williams New Series of Improved

DOUBLE ACTION

RootCutters

The "Triumph."

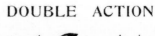

Fitted with Renewable Swivel Bearings.
                   Improved Agitator.
                   Capacious Hopper.
                   Detachable Bottom Hopper
                   Steel Legs
                   Unchokable Disc.

	T 7	T 8	
Pulper & Slicer.	24in. disc.	26in. disc.	If fitted with brass bearings,
Pulper & Finger-piece Cutter.			
Pulper & Grater.	£5 5 0	£5 12 6	
Grater & Slicer.			
Grater & Finger-piece Cutter.	£5 10 0	£5 17 6	7/6 extra.
Slicer & Finger-piece Cutter.	£5 10 0	£5 17 6	

Corbett-Williams

"New Patent"

RootCleaner

and Cutter.

"The Triumph"

Unique in Design.
    Unique in Action.
        A thorough cleaner.

UNCHOKABLE.

Takes little power.
Practical, Simple, Efficient.

The Cleaner is composed of a travelling web, the bars of which close together as they pass from the convex to the concave form, thus scraping the dirt from the root **without Chipping, Cutting, or Bruising** The fall of the roots from the Cleaner into the Hopper of the Cutter is regulated by means of a small wheel at the end of the machine.

PRICE—TC5. Pulper, Slicer, Grater, or Fingerer,   }
    26in. disc. and patent cleaner complete.   **£14 - 3 - 6**
To drive from either side, **31/6** extra.
In ordering machine with "Side Drive" please say if RIGHT or LEFT is required

Corbett-Williams
Single - Action Root Cutter

The "Climax."

26in. Disc, with Agitator.

	C I		
	£.	s.	d.
Pulper, with four chisel point knives	3	7	6
Pulper, with six chisel point knives	3	12	6
Pulper, with six diamond point knives	3	12	6
Grater, with gouge plate knives, coarse or fine cut ..	3	12	6
Slicer, with curved knives ..	3	12	6
Finger-piece Cutter ..	3	12	6

Corbett-Williams
Double-Action Root Cutter.

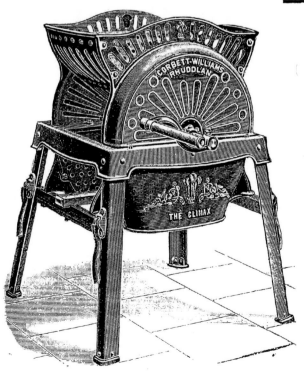

"The Climax."

26in. Disc, with Agitator and Steel Legs.

	C D A		
	£	s.	d.
Pulper and Slicer combined ..	5	5	0
Pulper and Finger Cutter combined..	5	5	0
Pulper and Grater combined ..	5	5	0
Slicer and Finger Cutter combined..	5	10	0
Grater and Slicer combined ..	5	10	0
Grater and Finger Cutter combined..	5	10	0

These Machines, which have been further improved, possess the same unique peculiarities of the Single-Action Machine on the preceding page, and are in workmanship, design, capacity and efficiency, the CHEAPEST AND BEST MACHINES ON THE MARKET.

Pulpers, if fitted with 6 knives instead of 4 knives, **5**/- extra.
If fitted with Brass Bearings, **7/6** extra.

Corbett-Williams

IMPROVED NEW PATENT

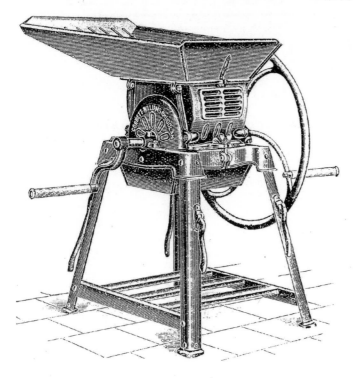

Gardner
Turnip Cutters

Single-action for Sheep, cuts ⅝ × ¾in.
Finger-pieces **£4 10 0**

Single-action for Cattle, cuts ⅝ × 1½in.
Finger-pieces **4 10 0**

Double-action for Sheep and Cattle,
cuts ⅝ × ¾in. Finger-pieces for Sheep,
and Slices ⅝in. thick for Cattle..   ..   **5 10 0**

Travelling Wheels, extra **10/-**

Boxed three sides, extra **7/6.**

This Machine can be supplied with a Wood
Frame and with Wood Legs, *or* with Wood Legs and
Iron Frame, at the same price.

Straight Front Plates to Hopper are supplied
unless ordered otherwise, but Curved Plates can
be supplied if desired.

Corbett-Williams

IMPROVED

Oil Cake Breaker.

PRICES.

No. 1, with Hopper 12 inches wide, will break six
different sizes. The Hopper can be enlarged
for breaking and accumulation of small pieces.
It is fitted with a Screen and Two Boxes for
receiving the crushed cake and dust separately.
Wheels enclosed by neat ornamental covers,
as shown on engraving   ..   **£3 6 0**

Without cover to gearing   ..   **£2 17 6**

No. 1 X, similar to No 1, but with 14 inch
Hopper **£3 10 0**

Without cover to gearing   ..   **£3 3 0**

No. 1 A, similar to No. 1 X, but with 15 inch
Hopper **£3 14 0**

Without cover to gearing   ..   **£3 6 0**

Corbett-Williams
NEW PATTERN IMPROVED
Corn and Malt Crusher.

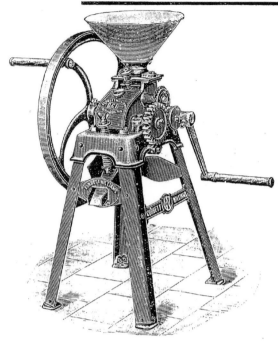

This design is the result of many years' experience in Grinding and Crushing Machinery, and embodies all the modern improvements.

The Main Frame, which is mounted on Rolled Steel Legs, is of Cast Iron, neat in design, but of good weight and ample strength. The main Bearings are fitted with Top and Bottom Brasses, and all Oil Holes are protected from dust by Brass Caps. A Parallel Motion Frame carries the Front Roller, and is adjusted by a Hand Wheel at the front, and it is also fitted with Safety Relief Spring, which prevents any hard foreign substance from damaging the rollers or gears. The Rollers are made of special hard close-grained Iron, and are accurately machined to size by special turning and fluting machinery. The Spindles throughout are of high-grade Steel.

DC1.	With Plain Rollers, 5in. × 5in.	£6	O	O
DC2.	With Finely Fluted Rollers, 5in. × 5in.	6	O	O
DC3.	With Coarsely Fluted Rollers, 5in. × 5in.	6	O	O

Corbett-Williams
Domestic Grinding Mills

For Hand Power.

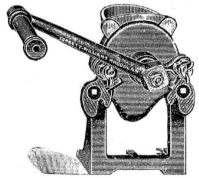

Mark: No. 0.—This Mill can be fixed to a rail or post, or fastened down on a table.

Price 15 /-

These Mills have been specially designed to meet the requirements of the Settler in the Colonies and Abroad, where it is necessary to grind grain fine enough for domestic purposes. They are suitable for Farms, Stables, Grocers, Chemists, or Poultry Feeders.

They will grind into fine, coarse, or moderately fine meal, or kibble all kinds of Grain, Coffee, Chicory, Cocoa, Charcoal, Drugs, Rice, Pepper, Spices Linseed, Wheat, Maize, Beans, Peas, Barley, Oats, Mealies, and will grind into the finest powder Chemical, Vegetable, or Mineral substances, broken Oyster Shells, Bones, Dog Biscuits, and Oil Cake.

Mark: No. 1.—Suitable for grinding into fine meal for domestic purposes.
Price £1 12 0

Mark: No. 2.—Suitable for Kibbling or Coarse Grinding.
Price £1 12 0

The above Mills can be fastened down on a Table or Bench.

Corbett-Williams Domestic Grinding Mills

For Hand Power.

Prices of Renewable Parts.

Mills marked Nos. 1, 2, 3, 4 :

Rotary Cone	.. ..	**1/6**
Rotary Plate	.. ..	**1/6**
Fixed Cone	.. ..	**2/6**

Mill marked No. 0 :

Rotary Plate	.. ..	**1/3**
Fixed Plate	.. ..	**1/3**

Mark : No. 3.

Suitable for grinding into fine meal for domestic purposes.

Price **£2 7 0**

———

Mark : No. 4.

Suitable for kibbling or coarse grinding.

Price **£2 7 0**

———

Mounted on Cast Iron Table with Steel Legs.

Mills Nos. 1, 2, 3, and 4 are fitted with Self-adjusting Grinding Plates, which ensure their running perfectly true one with the other, thus guaranteeing a good even sample of grinding. All wearing parts and grinding plates can be easily replaced when worn. The above Mills are specially designed for export, and can be packed in a very small space for shipment in quantities of six or twelve. If twelve Mills are packed in a case, the cost of packing is considerably reduced.

CORBETT-WILLIAMS
Improved New Pattern Safety Horse Gears,

WITH 30in. and 36in. DRIVING WHEELS. LATEST DESIGNS.

These Gears are newly designed, and are fitted with Dome Wheels, bridged over by a substantial frame firmly braced together, renewable chilled Toe Steps for upright shafts. Safety Clutch, so that the machinery will run should the horse come to a stand or slacken his pace. The Gears are well adapted for driving Churns, Pumps, Stacking Machines, Chaff-cutters, and other Barn Implements.

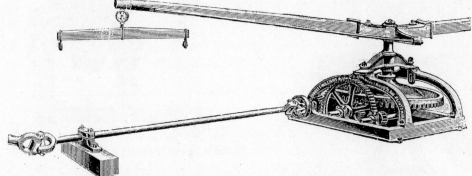

No. 2.—Pony or Mule, with 30in. Driving Wheel and Brass Bearings, and Reversible Safety Clutch. PRICE, complete for One Pony and fitted with two speeds. 6 and 20 revolutions to one of the Pony **£7 15 0**

No. 2A.—With two Speeds, 6 and 36 revolutions to one of the Pony **7 15 0**

No. 2B.—With three Speeds, 6, 20, and 36 revolutions to one of the Pony **8 5 0**

No. 3.—Horse Gear, with 36in. Driving Wheel, Brass Bearings, and Reversible Safety Clutch. PRICE, complete for One Horse and fitted with three Speeds, 6, 20, and 36 revolutions to one of the Horse **12 0 0**

PRICE, complete for Two Horses **13 0 0**

One-horse Gears are fitted with a Single Pole Socket, but can be fitted with a Double Pole Socket (as illustrated) and One Pole only at **5/-** extra.

ABSOLUTELY THE BEST VALUE ON THE MARKET.

CORBETT-WILLIAMS
Improved New Pattern Safety Horse Gears,

WITH 30-in. and 36-in. DRIVING WHEELS.   LATEST DESIGNS.

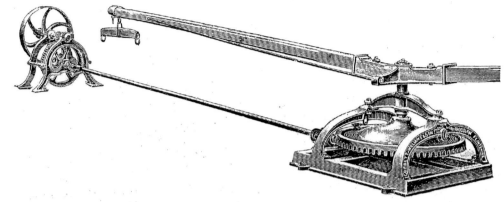

No. 5.—**Light One Horse Gear,** 30in. driving wheel ; speed, six revolutions to one of the horse.. **£6 15 0**
    Separate Intermediate Motion, with renewable bearings, increasing speed 25, 30, or
      36 revolutions to one of the horse **2 10 0**

No. 6.—**One Horse Gear,** 36in. driving wheel ; speed, six revolutions to one of the horse .. **9 12 6**
    Price complete for Two Horses **10 12 6**
    Separate Intermediate Motion, with renewable bearings, increasing speed to 25, 30,
      or 36 revolutions to one of the horse..   ..   .. **3 0 0**

Intermediate Motions are fitted with renewable Cast Bearings, unless ordered to the contrary. If supplied
with Brass Bearings to Top Spindle, **5/-** extra ; if supplied with Brass Bearings to Top and Bottom
Spindles, **10/-** extra.

Intermediate Motions can be supplied with Coupling and Short Connecting Rod, if required, instead of pulley.
One-horse Gears are fitted with a Single-pole Socket, but can be fitted with a Double-pole Socket and One Pole
only at **5/-** extra.

STRONG, SERVICEABLE, AND EASY RUNNING.

Corbett-Williams Oil Engines

THE

MOST

RELIABLE,

ECONOMICAL,

AND

SIMPLE

ENGINE

MADE.

PARTICULARS

OF THE

"Corbett-Williams" Horizontal Oil Engines.

ORDINARY TYPE FOR REFINED OIL (PETROLEUM, PARAFFIN, or KEROSENE.)

MAXIMUM. B H.P.	Net Weight Appr'x	Gross Weight Packed for Shipment Appr'x	Approximate Cubic Measurements: Without Tank and Pipes	Approximate Cubic Measurements: Complete	Space occupied by Engine only	Net Weight of Standard Portable Engines	Approximate Cubic Measurements Standard Portable Engines	Standard Size of Pulley	Size of Fly-wheels	Speed Revolutions per Minute	MAXIMUM. B.H.P.	PRICES. Without Tank and Pipes	PRICES. Complete with Tank and Pipes	PRICES. Standard Portable
	Cwts	Cwts	Cub. ft	Cub. ft.	Ft. Ins. Ft. Ins.	Cwts	Cub. ft.	Ins.	Ft. Ins			£	£	£
4	10	13	34	55	4 9 x 2 8	23	190	10 x 6	2 6	360	4			
4½	10½	13½	38	59	4 10 x 2 9	23½	200	10 x 6	2 6	360	4½			
5½	14	17	40	60	5 1 x 3 4	29	210	12 x 6	2 8	350	5½			
6½	15	18	45	65	5 3 x 3 6	30	220	12 x 6	2 10	340	6½			
8½	22	28	60	100	6 2 x 4 5	35	220	16 x 9	3 6	320	8½			
10	22½	29	65	105	6 4 x 4 5	37	225	16 x 9	3 6	300	10			
11½	23	30	70	110	6 6 x 4 6	39	230	16 x 9	3 8	280	11½			
13½	28	35	75	115	6 8 x 4 7	45	240	18 x 10	3 8	275	13½			
15½	36	44	100	176	7 3 x 4 8	52	320	20 x 12	4 2	260	15½			
17	40	49	110	186	7 4 x 4 10	56	350	20 x 12	4 2	250	17			
19	44	53	132	208	8 1 x 5 7	60	380	20 x 12	4 4	250	19			
21	57	67	134	210	8 2 x 5 8	70	460	22 x 12	4 6	240	21			
23½	60	71	183	263	8 5 x 6 0	72	480	22 x 12	4 6	240	23½			
26½	63	75	210	290	8 9 x 6 1	75	500	24 x 13	4 9	240	26½			

All Engines are fitted with Two Fly-wheels.

ALL MEASUREMENTS, DIMENSIONS, AND WEIGHTS GIVEN ARE APPROXIMATE.

"Corbett-Williams" Vertical Petrol Engines

3 H.P. and 5 H.P.

Starts instantly. No lamp required. Complete with High Tension Magneto Ignition, Water Tank and Standard Connections, Silencer and Standard Pulley.

Stationary Type.

Brake H.P.	Speed R.P.M.	Flywheel diam.	Driving Pulley.	PRICE.
3	450	22″	8″ x 5″	£
5	450	24″	10″ x 6″	£

"Corbett-Williams" Portable Petrol Engines.

Mounted on Strong 4-wheeled Wooden Trolley with Cooling Tank.

Portable Type.

Brake H.P.	Speed. R.P.M.	Flywheel diam.	Driving Pulley.	Price with Ordinary Standard Cooling Tank.	Price with Square Cooling Tank and Circulating Pump.
3	450	22″	8″ x 5″	£	£
5	450	24″	10″ x 6″	£	£

N.B.—The Square Cooling Tank and Circulating Pump are lighter for transport than the Ordinary Cooling Tank, and, moreover, where there is a limited water supply this arrangement is a distinct advantage.

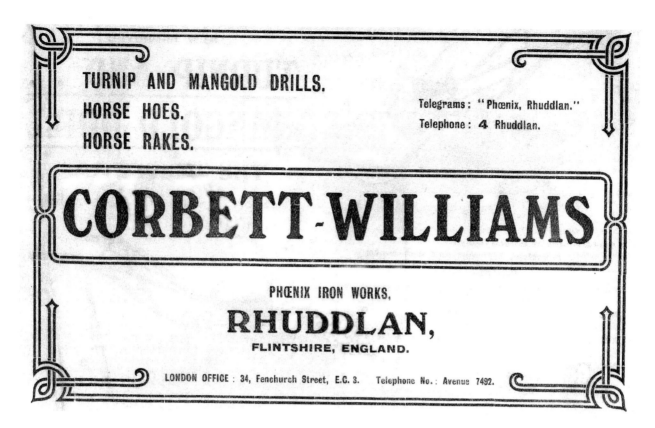
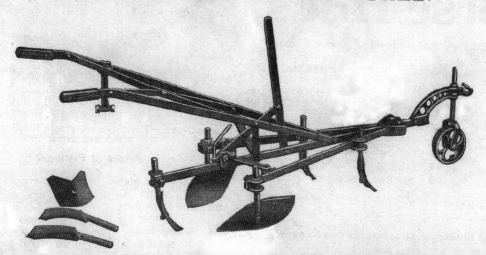

Corbett-Williams
New Pattern Expanding Horse Hoe

By taking off the two Side Hoes and substituting the two extra Tines as shown in the illustration on preceding page, the Hoe can be converted into a 5-Tine Cultivator capable of cultivating up to a width of 20 inches. The 5-Tines can be arranged at distances of 5 inches apart, and, by moving the adjusting lever, they can be regulated to cultivate to any width under 20 inches wide, and at any required distance apart. Also the Tines can be arranged so that they miss no ground when cultivating. Thus, the 5-Tines, which are 2½ inches wide, could, if required, cultivate a width of less than 10 inches, thus thoroughly pulverising and digging the soil the whole depth of the Tine.

The two Front Tines fitted on the two Side Bars can be adjusted longitudinally, and can be brought very close together by fixing same on the inside of the Bars instead of on the outside.

The depth of the soil required to be cultivated can be arranged by altering the height of the Wheel at the front, and the Tines can also be adjusted with a Set-screw.

The Side Hoes being fitted with round Stalks, can be reversed with the sweep outside for earthing up when required.

The Shovel shown in the illustrations can be fitted to the front or back of the Hoe as desired.

All expanding Hoes are supplied with the same complete equipments, whereby the Implement can be fitted up in either style as shown in the three illustrations.

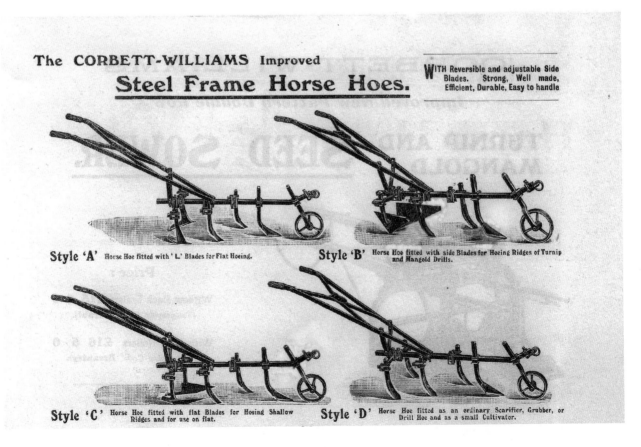

The CORBETT-WILLIAMS Improved
Steel Frame Horse Hoes.

WITH Reversible and adjustable Side Blades. Strong, Well made, Efficient, Durable, Easy to handle

Style 'A' Horse Hoe fitted with 'L' Blades for Flat Hoeing.

Style 'B' Horse Hoe fitted with side Blades for Hoeing Ridges of Turnip and Mangold Drills.

Style 'C' Horse Hoe fitted with flat Blades for Hoeing Shallow Ridges and for use on flat.

Style 'D' Horse Hoe fitted as an ordinary Scarifier, Grubber, or Drill Hoe and as a small Cultivator.

Corbett-Williams

IMPROVED STEEL FRAME

Horse Hoes

Malleable Castings used throughout on these Hoes.

Strong, Well Made, Efficient, Durable, Easy to Handle.

MADE IN FOUR SIZES TO SUIT ALL SOILS.

PRICES :
No. 1 (For Market Gardeners)	£2 17	6
„ 2 (Medium)	£3 3	0
„ 3 (Strong)..	£3 5	0
„ 4 (Extra Strong).. ..	£3 8	0

When ordering please state which style of Hoe is required, namely, Style A, B, C, or D.

Any additional Blades or attachments can be supplied at Prices shown.

An extra pair of Reversible Points is included in the prices shown

A ridging body can easily be fitted on at back in the socket which is provided on each Hoe.

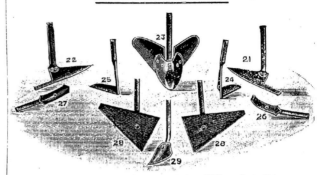
Awarded First Prize and Special Bronze Medal at the Allahabad (India) Exhibition, held 1910-1911.

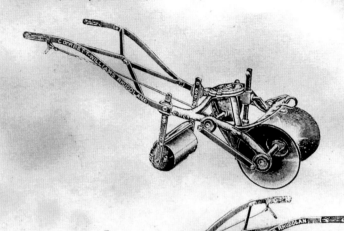

The Corbett-Williams
NEW IMPROVED
Turnip and
Mangold Drill

The "New Era."

STEEL SPINDLES AND FRAMING THROUGHOUT.

Special Note to Colonial and Foreign Buyers.

Where it is the custom to sow on the flat instead of on the ridge, we supply this Drill fitted with two wheels in place of the front concave roller, and with the flat roller for covering the seed fitted at back.

Price £2 - 0 - 0

A New Machine of Unique Design—Practical, Efficient, and Well-made.

Price—				
With One Concave Roller	£2 0 0	..	(Telegraphic Code—Era).
With Wheels in lieu of Front Roller..	£2 0 0	..	(Telegraphic Code—Erasmic).
With Two Concave Rollers ..	2 8 6	..	(Telegraphic Code—Erat).
With Wheels in lieu of Rear Roller	2 8 6	..	(Telegraphic Code—Erratic).

This Machine is provided with a Handpiece under the Handle, so that the seed can be instantly stopped when turning at the headlands. The driving cranks are fixed on the spindles with cross pins, and cannot work loose, no nuts being used.

Corbett-Williams

IMPROVED

NEW PATTERN

DOUBLE ROW

TURNIP AND
MANGOLD

SEED -
SOWER.

Price :

Without Back Rollers .. **£6 0 0**
(Telegraphic Code—Errant).

With Back Rollers **£6 10 0**
(Telegraphic Code—Errantry).

Corbett-Williams
Single Row Mealie Planters

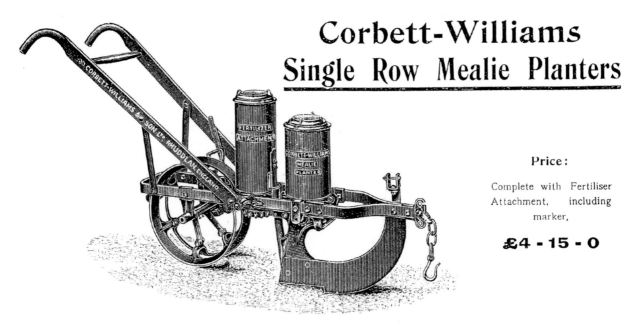

Price:

Complete with Fertiliser Attachment, including marker,

£4 - 15 - 0

Corbett, Williams & Son, Ltd., having had a large experience (extending over sixty years) in the Manufacture of Machines for Planting Seeds of all descriptions, have devoted much time and labour in introducing the above Mealie Planter, which will be found a most effective, reliable, and simple implement. The Planter has been designed and introduced specially to meet the requirements of the Colonial Planter.

The Machine is furnished with a complete set of Delivery Plates, with holes suitable for all grades of Maize, and also with Two Plates for Kaffir Corn and other small seeds. Each section of plates has a different number of holes, so that by changing the plates the distance between the sets can be varied as desired. This compact Planter will deposit Maize at various distances apart, and it will also plant Beans, Peas, Cotton, and other seeds at suitable distances. It is fitted with a marker to enable the operator to keep the rows at equal distances apart. The depth is regulated by the hitch on the head of the implement. The handles are adjustable in height. Planting is started or stopped by a lever, by means of which the machine can be put in and out of gear. The illustration above shows the Planter fitted with an open concave wheel, by which the seeds are effectually covered by the soil closing over them. If desired, a closed concave wheel can be supplied instead of the open concave wheel, but unless otherwise ordered, the open concave wheel (as illustrated) will always be supplied.

CORBETT-WILLIAMS
ALL STEEL IMPROVED THROUGH-AXLE # MORSE RAKE.

Easy
Adjustment.

Perfect
Balance.

Adjustable
Seat.

Light
Leverage.

MANUAL AND FOOT LEVER.

No.	Teeth	Height of Wheels.	Width.	Code Word.		Price		
R1.	24	... 52in. ...	7ft. 8in. ...	Anchor	...	£25	10	0
R2.	26	... 52in. ...	8ft. 2in. ...	Angel	...	£25	15	0
R3.	28	... 52in. ...	8ft. 8in. ...	Anvil	...	£26	0	0

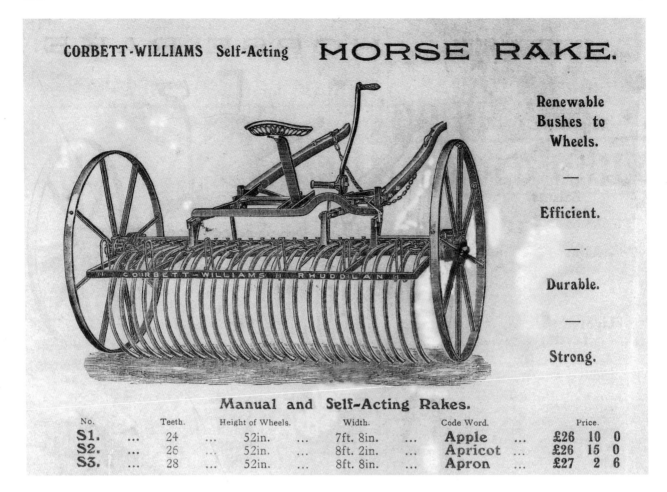

CORBETT-WILLIAMS Self-Acting **MORSE RAKE.**

Renewable
Bushes to
Wheels.

—

Efficient.

—

Durable.

—

Strong.

Manual and Self-Acting Rakes.

No.		Teeth.		Height of Wheels.		Width.		Code Word.		Price.		
S1.	...	24	...	52in.	...	7ft. 8in.	...	Apple	...	£26	10	0
S2.	...	26	...	52in.	...	8ft. 2in.	...	Apricot	...	£26	15	0
S3.	...	28	...	52in.	...	8ft. 8in.	...	Apron	...	£27	2	6

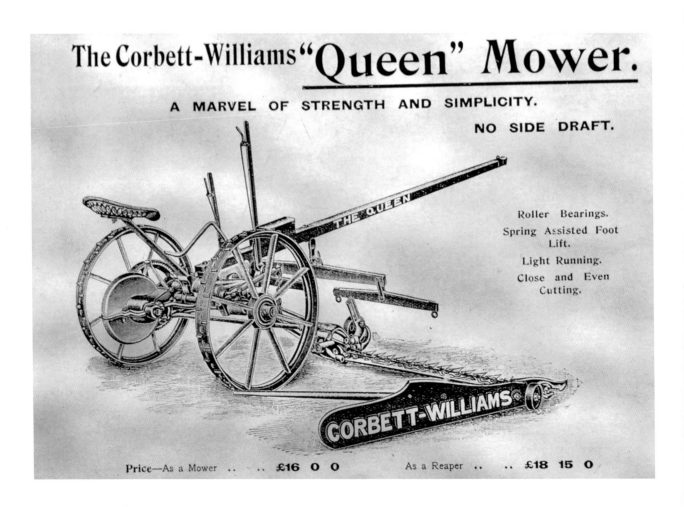

The Corbett-Williams "Queen" Mower.

A MARVEL OF STRENGTH AND SIMPLICITY.

NO SIDE DRAFT.

Roller Bearings.

Spring Assisted Foot
Lift.

Light Running.

Close and Even
Cutting.

Price—As a Mower	..	..	£16 0 0	As a Reaper	..	..	£18 15 0

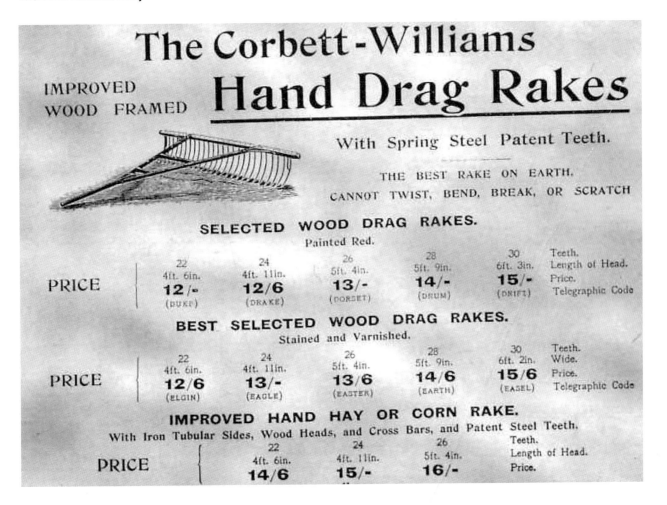

The Corbett-Williams

IMPROVED WOOD FRAMED

Hand Drag Rakes

With Spring Steel Patent Teeth.

THE BEST RAKE ON EARTH.
CANNOT TWIST, BEND, BREAK, OR SCRATCH

SELECTED WOOD DRAG RAKES.

Painted Red.

| PRICE | 22 4ft. 6in. 12/- (DUKE) | 24 4ft. 11in. 12/6 (DRAKE) | 26 5ft. 4in. 13/- (DORSET) | 28 5ft. 9in. 14/- (DRUM) | 30 6ft. 3in. 15/- (DRIFT) | Teeth. Length of Head. Price. Telegraphic Code |

BEST SELECTED WOOD DRAG RAKES.

Stained and Varnished.

| PRICE | 22 4ft. 6in. 12/6 (ELGIN) | 24 4ft. 11in. 13/- (EAGLE) | 26 5ft. 4in. 13/6 (EASTER) | 28 5ft. 9in. 14/6 (EARTH) | 30 6ft. 2in. 15/6 (EASEL) | Teeth. Wide. Price. Telegraphic Code |

IMPROVED HAND HAY OR CORN RAKE.

With Iron Tubular Sides, Wood Heads, and Cross Bars, and Patent Steel Teeth.

| PRICE | 22 4ft. 6in. 14/6 | 24 4ft. 11in. 15/- | 26 5ft. 4in. 16/- | Teeth. Length of Head. Price. |

Corbett-Williams

NEW SEGMENT PATTERN

FIELD ROLLER.

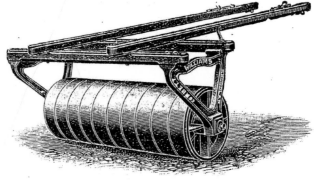

This implement is composed of a series of rings with flat surfaces, which, working independently on a round spindle, adjust themselves to the undulating surfaces much better than the ordinary barrel roller, and is more easily turned at the ends of fields. They are fitted with Balanced End Brackets and renewable Bearings, Wood or Tubular Frame and Shafts.

Can be fitted with special pole for bullocks or oxen, at same prices.

SEGMENT LAND ROLLERS.

Width. Ft. in.	16in. dia. £ s. d.	18in. dia. £ s. d.	20in. dia. £ s. d.	22in. dia. £ s. d.	24in. dia. £ s. d.	26in. dia. £ s. d.
5 6	9 0 0	10 0 0	11 0 0	11 10 0	12 5 0	13 0 0
6 0	9 10 0	10 10 0	11 5 0	11 15 0	12 10 0	13 10 0
6 6	10 0 0	11 0 0	11 15 0	12 10 0	13 5 0	14 10 0
7 0	10 10 0	11 10 0	12 10 0	13 5 0	14 0 0	15 15 0

Corbett-Williams
IMPROVED CAMBRIDGE PATTERN

CLOD - -
CRUSHER.

These Clod Crushers are composed of a number of wheels, 3 inches wide, with thin cutting edges, each turning freely on the spindle. They are fitted with Balanced End Brackets and renewable Bearings, Wood or Tubular Frame and Shafts.

Can be fitted with special pole for bullocks or oxen, at same prices.

CAMBRIDGE ROLLERS.

Width. Ft. in.	16in. dia. £ s. d.	18in. dia. £ s. d.	20in. dia. £ s. d.	22in. dia. £ s. d.	24in. dia. £ s. d.	26in. dia. £ s. d.
5 6	9 10 0	10 0 0	10 15 0	11 5 0	12 0 0	15 0 0
6 0	9 15 0	10 5 0	11 0 0	11 10 0	12 10 0	15 15 0
6 6	10 5 0	10 15 0	11 10 0	12 10 0	13 10 0	16 10 0
7 0	10 15 0	11 5 0	12 0 0	13 0 0	14 5 0	17 5 0

If fitted with 2in. Rings **20/-** extra.   Driver's Seat, **15/-** extra.

Patent "Phœnix" Harrows

FOR THE FOREIGN AND COLONIAL MARKETS.

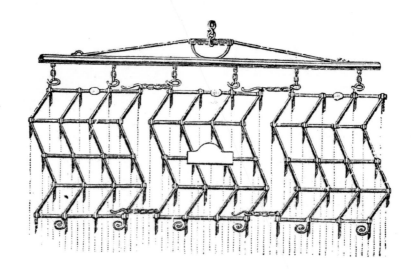

Upwards of One Hundred Thousand of these Harrows are now in use in all parts of the world, and are giving universal satisfaction.

Corbett-Williams
Cast Iron Pulleys.

Diameter in Inches.	PRICE OF BLACK PULLEYS.		PRICE OF TURNED PULLEYS.	
	3¾in. wide.	6in. wide.	3½in. wide.	5½in. wide.
8	4/-	5/-	5/6	8/-
9	4/6	5/6	6/3	9/-
10	5/-	6/3	7/-	10/-
11	5/6	6/9	7/9	11/-
12	6/.	7/6	8/6	12/-
13	6/6	8/-	9/3	13/-
14	7/-	8/9	10/-	14/-
15	7/6	9/3	10/9	15/-
16	8/-	10/-	11/6	16/-
17	8/6	10/9	12/3	17/-
18	9/-	11/6	13/-	18/-
19	9/6	12/3	13/9	19/-
20	10/-	13/-	14/6	20/-
22	11/-	14/-	16/-	22/-
24	12/-	15/-	17/6	24/-
26	—	17/6	—	27/-
28	—	20/-	—	30/-
30	—	22/6	—	33/-

Chaff-Cutter Knives.

Nos. F1, F2, F3, F4, F5, AB, AC, AD, ADA, AE, and WPI Machines	5/- per pair.
„ WP1A and WP2 Machines	7/6 „
„ WP3 Machine	8/6 „
„ WP4	8/6 „
„ WP5	10/- „
„ WP3A, WP4A (Set of Three)	15/- per set.
„ WP6 „	15/- „
„ WP7 „	18/- „

INDENTURE

1920

First day of September One thousand nine hundred and
twenty BETWEEN Corbett, Williams & Son Limited, of Rhuddlan
in the County of Flint, Agricultural and General Engineers,
Machinists, and Manufacturers of Internal Combustion
Engines, hereinafter called the said Company of the first
part, David Johnson Humphreys, Son of Thomas Johnson
Humphreys of Dyserth, in the County of Flint, hereinafter
called the said Apprentice of the second part and the
said Thomas Johnson Humphreys hereinafter called the
Parent of the third part WITNESSETH, that in consideration
of the good and faithful service of the said David Johnson
Humphreys, the Apprentice, to be done and performed to
or for the said Company, and in consideration of the sum of
One hundred and fifty pounds paid by the said Parent to the
Company, the receipt of which the Company hereby acknow-
ledges, and of the covenants and agreements hereinafter
entered into by the said Apprentice and the said Parent,
the said Company at the request of the said Parent and with
the consent of the said Apprentice testified by his executing
these presents DOTH HEREBY COVENANT AND AGREE with the said
Parent and also with the said Apprentice in manner following
that is to say:- THAT the said Company will take and receive
the said Apprentice as their Apprentice from the first day
of September one thousand nine hundred and twenty for the
term of four years and seven months, AND ALSO will during
the said term to the best of their knowledge power and
ability instruct or cause to be instructed the said
Apprentice in the trade of Agricultural and General
Engineers, Machinists and Manufacturers of Internal
Combustion Engines in such manner as the said Company do
now or shall hereafter during the said term use or practise
the same AND ALSO will weekly and every week during the said
term pay to the said Apprentice during such time only as he
shall actually work and be employed as such Apprentice in
such trade as aforesaid wages according to the following
scale that is to say :-
DURING THE FIRST YEAR THE SUM OF EIGHT SHILLINGS PER WEEK
during the second year the sum of nine shillings per week
during the third year the sum of ten shillings per week
during the fourth year the sum of twelve shillings per week
during the fifth year the sum of fifteen shillings per week
AND in consideration of the covenants and agreements
hereinbefore contained on the part of the said Company
the said Parent doth hereby place and bind the said
Apprentice and the said Apprentice with the consent and

approval of the said parent doth hereby place and bind
himself with and to the said Company for and during the
term aforesaid during all which time the said Apprentice
shall faithfully diligently and honestly serve the said
Company and obey and perform all their lawful and
reasonable commands and requirements and those of the
Foreman Heads of Departments or Instructors under whom
the said Apprentice shall be placed and shall not do any
damage or injury to the said Company or the property
of the said Company or knowingly suffer the same to be
done by others without acquainting the said Company or
their Foreman Heads of Departments or Instructors there-
with but shall in all things conduct and acquit himself
as an honest and faithful Apprentice ought to do, AND
for the consideration aforesaid the said Parent doth
hereby covenant and agree with the said Company that the
said Apprentice shall faithfully honestly and diligently
serve the said Company or their Foreman Heads of
Departments and Instructors and shall not waste embezzle
give or lend any of the money or property of the said
Company which may be in his possession or under his
control AND ALSO that the said Parent his executors or
administrators will at their own expense find and provide
the said Apprentice with good and sufficient board,
lodging, clothing, washing, pocket money, medicine, and
medical attendance and all other necessaries during the
said term AND ALSO that the said Parent his executors or
administrators will indemnify and keep indemnified the
said Company against any loss which the said Company may
sustain by reason of any embezzlement or other wrongful
act or default of the said Apprentice during the said
term PROVIDED ALSO AND IT IS HEREBY EXPRESSLY AGREED that
in case the said Apprentice shall at any time during the
said term be wilfully disobedient to the said Company or
to any Foreman or Instructor under whom by the commands
of the said Company he is placed or shall absent himself
without leave or lawful excuse or shall otherwise grossly
misconduct himself it shall be lawful for the said
Company to discharge the said Apprentice from their
service and if any money shall be due to him for wages
under this INDENTURE then in the case of such dismissal
the same shall be actually forfeited.   PROVIDED ALWAYS
AND IT IS HEREBY EXPRESSLY AGREED between the said parties
hereto that the said Apprentice shall not be or become a
member of any Trade Society or Trades Union or take part
in any trade dispute during the continuance of his term
of Apprenticeship unless with the consent in writing of
the said Company or their Managing Director or Directors
first had and obtained AND IT IS HEREBY AGREED that the
said Apprentice shall work such overtime as may be
reasonably and legally required by the said Company or
their Managing Director or Directors and he shall be
paid for the same at the rate allowed to Apprentices by
the said Company for such overtime work in accordance

the rules and regulations of the said Company in force
at the time of signing this Indenture.

IN WITNESS whereof the said parties have hereunto set
their hands the day and year first above written

SIGNED by the said Corbett,   ) **For CORBETT, WILLIAMS & SON, LIMITED**
Williams & Son Ltd in the   )
presence of :-

A.J.Green
Rhiouva
Rhuddlan

                                           WORKS DIRECTOR.

SIGNED by the said David   ) *David J Humphreys*
Johnson Humphreys in the   )
presence of :-   )

Howard. J. O. Lewis
Oaklea
Rhuddlan

SIGNED by the said Thomas   ) *T J Humphreys*
Johnson Humphreys in the   )
presence of :-   )

Howard. J. O. Lewis
Oaklea
Rhuddlan

Bibliography

A Directory of Agricultural Machinery and Implement Makers in Wales. Elfyn Scourfield National Museum of Wales.

Agricultural Implement Makers John Williams & Son Rhuddlan. Messers Powell Brothers & Whitaker, Wrexham. Elfyn Scourfield National Museum of Wales 1983.; Revised version of two articles reproduced by kind permission of their respective editors.

'John Williams & Son Phoenix Ironworks, Rhuddlan' in Flintshire Historical Society Journal, Volume 28, 1977-1978. Elfyn Scourfield.

'Messers Powell Brothers & Whitaker Implement Makers, Wrexham' in Denbighshire Historical Society Transactions, Volume 25, 1976. Elfyn Scourfield.

Maritime History of Rhyl and Rhuddlan D.W. Harris 1991.

The Vale of Clwyd Railway Rhyl to Denbigh Stephen P. Goodall 1992.

IRON HARVESTS OF THE FIELD The Making of Farm Machinery in Britain since 1800. Peter Dewey 2008.